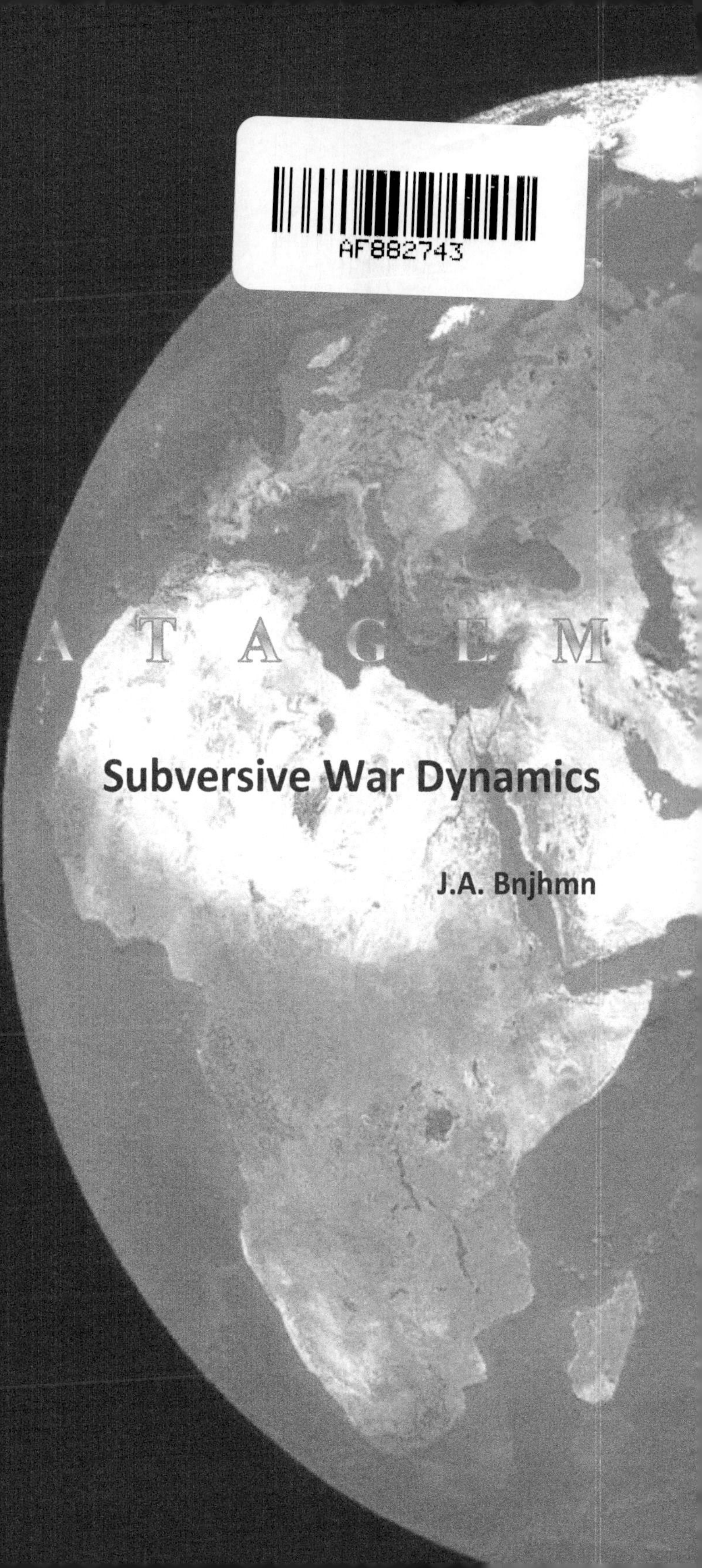

STRATAGEM

Subversive War Dynamics

J.A. Bnjhmn

CONTENT

One’s Ignorance is Another’s Omnipotence

Introduction

Interaction and perception are weapons paramount in the assault fashioning a mind-set imperceptible, even to those oppose. Of the contemporary for example the psychological ploy 'terrorist platform' utilized to affect paranoia, suspicion, influences anarchist behavior of host mass populace, the manner which the system dictates (prescribe requirement) affects psychosis. Propaganda tool allows for the enforcement of draconian authoritative measure providing protective shield for those devising contradictory global chaos. This is psychology of such magnitude that many affected thereby do not detect its influence. Tool of propaganda is undeniably the molder of the very image of global humanity. Media, the potter of the clay mind, influence, imposes, suggest, sometime instigates. The producers and directors of this production number have been quite busy introducing to the global mind-set a bevy of stage, phase theatrics. This type of imaginary drama affects the very image, character of culture.

Psychology has cultivated a society of race, class, cult faction. This division bell assures chaos, confusion provide for the 'playing of both sides between the middle' control mechanism. It is nothing more than cultural-ethnic competition exacerbated for the purpose of social engineering, resource acquisition and control. Most conflicts are orchestrated of provocateurs less aware the turmoil. The psycho socio-economic imperative is the dynamic by which destabilizing initiative are affected; OBJECTIVES ACHEIVED. War stratagem *is* the constant, utilizing various elements in a cultural mainframe to achieve objective; with the pretense of peace projected, ever so expressed.

Man's perception of self has exacted the proper denouncement for his existence. Society, a diverse mix confers upon humanity unique identities, characteristic, physical traits of a primal origin. Global impetus affects to set the many parts of the amalgam into acceptable psychological collective mind-set rendering identity, ethnicity a mundane tribal attribute. Social engineering is a dynamic of dictate which confers on the architect exclusive right of inclusion. Mutual agreement on matters of civility, acceptance, are noble at best, but crude dictates demonstrate double-standard where identity renders collectivity a moot point.

Prologue

Divide and conquer has a new mantra; destabilize, realign and control. Conventional weaponry has gone the way of occurrence, events to affect accorded response. Choice weaponry speaks of propaganda, strife, need where a global population of seven (7) billion plus must concern itself with the trial and tribulation of survival no less the ancestral cave man. One can observe this phenomenon on a daily basis whether in so-called third-world setting or navigating their industrialized culture. Hunt and gather requires you find a job, earn a wage, pay the bill. Many have preferred the laws of nature to the edict of capitalism. Global edict would suggest otherwise.

This is not the "land of the free" proclamation that spearheaded a nation but a bold collective global scheme set on relegating the mass populace to the world order scheme you now see playing out before you. Destabilization is a war stratagem (a contemporary assault on cultural religion) imposing and has divested of all other perception. Conventional barbarity does not allow for spiritual reticence where Western faith relies on weaponry and the province of...annihilation. To achieve these objectives ethical standard has been giving 'means necessary' makeover. This is evident of various market collapses requiring a socialistic resolve. Draconian effect and fascist curtailment of civil liberties, where the argument posited espouses the authoritarian, despotic. Fraud is the acceptable business acumen, whereas terms terrorism and transparency provide cover for deviate, corrupt practice.

In this millennial, conspiracy has provided for debate of cause and effect. Deliberate action dictates chaos, crisis management orchestrated by American, U.N. forces hell-bent on bringing about the desired result. Various military occupations that have occurred (Iraq being the litmus) were initiated of pure speculation rendering the actions murderous, barbaric. The boast of patriotism now carries a proposal decreeing death pronounced on those espousing a religious preference, difference of opinion in the very same manner as a supposed enemy. The bold scheme dictate hostile action includes all.

The world order scheme is a timeline of propose stage phase event. Although the plot is of ancient, contemporary trends politics, religion, class, race, status is in process of rework. Symbolic gesture, technological innovation has become the rallying cry for all to hop on board. As the script unfolds the detrimental effects devastate the whole of humanity. It must be understood the sinister scheme is older than every being living in the present, enliven thereafter all seven billion have expired. Time-line event are orchestrated to bring about desired effect.

What gives rise to event is *necessity* (defined as imperative) to achieve objectives purposed for desired outcome. Primal elements utilized to catalyze incident are of the psychological (mind), social (human interaction), economic (business) dynamic correlative of global disparities. Global humanity has reached a similitude of religious preference which is being persecuted for it obstructs the way of new world edict.

The new world order initiative began with the spurring effects of political heavies and watchdogs decrying the effects of big government (which is the welfare, entitlement program of the rich) and the need to curtail its regulatory influence. Curtailing or eliminating regulatory process began the circumvention of constitutional provision. Supply-side economics, Reagan-Era acts were embraced of big business giving them global expansionist preferences in the matter of relocating jobs and resources (outsourcing) throughout the international community where labor cost were substantially minimized as to the margin of profit. It presented grand opportunity to spearhead the cultivation of the global scheme. A compilation of the dynamic provides for definitive fundamental study, assessment.

Author's Note to the Reader

Utilizing Media Reference...has become somewhat a circumspect cause. Global news conglomerates producing commercialized medium offer pretentious, fabricated occurrence distorting subjective content for the effect of swaying albeit skewing perception to influence favor.

Content cannot be deemed evidentiary of the source event requiring critical assessment. Power elite manipulate tool of information for the purpose of propagating spin, conjecture to affect favorable opinion, misinforming for the sake of deception. This has made the profession of research precarious (writing from a western perspective) to say the least. In the age information technology 'transparency' can be considered no more than aberration. Critical analysis, skepticism should be well put to extraneous utility. Observation holds even with this writing the reader be critical of the writer.

Subversion

Demoralize, Destabilize, Conflict, Control

During the Cold War Era, 'diplomatic intrigue,' 'espionage,' were the terms used to describe the covert acts of international rivals infiltrating their adversaries social construct to undermine, influence disruption or control, in most cases, the governing body. Embassies and ambassadors were the actors conferred the diplomatic connect of such alleged action, resulting in their expulsion of such under legislative protection of 'diplomatic immunity.'

Today, the embassy is a Non-Government Organization (NGO), ambassadors are directors, corporate heads of such organizations. Intrigue and espionage have been replaced of terms such as 'initiative', and 'democracy.' It is a reconceptualization of the old intrinsic war mode with a 'psychological twist' laying the broader action of aggression on the adversary. This is war of subversive influence, instigation, incitement, conflict, from within; being least aggressive only when objectives call therefore.

Achieving such 'objectives' call for a thorough analysis of the adverse host cultural complexities allowing for the execution and precipitation of caustic effects to occur, proliferate. It is of a truth, the 'process psychological' goes undetected (sometimes by way of indifference) by host majority and affirms the deliberate intent.

Diplomacy and the perks (aid of some sort) that go therewith considered in the international arena, 'softpower' describe as the ability to attract and co-opt rather than coerce, use force or [as to] give money as a means of 'persuasion;' the empirical element of subversion. Excerpt of recent article espousing U.S. – Pakistani relation give prescient to the utility of such subversive diplomacy.

"American readiness to offer [aid] has bred dependence and hubris. The U.S. has ended up as an 'enabler' of Pakistan's 'dysfunction' by reinforcing the belief of its elite."[1]

Consternate of Pakistan political, security woes pandering to dysfunctional administration exacerbate the lack of conscionable consideration as the following illustrates;

"The operations that resulted in Bin Laden's death took place [without] any prior notification, explanation, or cooperation with the Pakistani authorities. This incident closely followed a CIA drone attack on Pakistani soil in March 2011, which was considered one of the deadliest ever; in addition to the Raymond Davis case, in which a US citizen, later revealed to be a CIA contractor, shot and killed two Pakistani in Lahore."[2]

---

[1] Can We Trust Our 'Frenemy' Pakistan? – Washington Examiner, Nov. 2 2015

[2] Revisiting China's 'String of Pearls' Strategy: Places 'with Chinese Characteristics' and their Security Implications – V. Marantidou, Issue & Insights, June 2014 Vol. 14 No. 7 pg. 12

These acts constitute a manner of conduct disconcerting, potentially catastrophic observed of a broader international audience illuminating a character contradictory. Such diplomacy is a double-edge sword perspective of dire consequence. Both actors of this prescriptive game of subversive political economic relation are complicit of the other action, inaction. 'Looking the other way,' speaks of withdraw from sovereign principle; undermining. For the sake of foreign aid (United States Pakistan largest trading partner) harboring terrorist, nuclear proliferation keeps Pakistan well poised to maintain Washington's attention both persuasively and coercively. This manner of conduct has become objective of the geopolitical war strategist.

Demoralization

'to deprive person of spirit, courage or discipline; to reduce to a state of weakness, disorder.'

Humanity has evolved into a constructive social dependency allowing for influences which dictate course action. It is imperative of the ruling class (global) they maintain or at least direct the perspective whole of class, culture. The psychology of perception has been a formidable weapon in the hands of those capable of manipulation. Subversive technique requires a host, cultural base, an assessment of the best methods suitable for the impending psychological assault. This is war of influence, manipulation to depose a designated adversary or host by means of eroding confidence, morale; defying spirit. These are precursor which fuel purpose, determination, drive. They set the stage for endurance, accomplishment. Demoralization however, prescribes defeat.

Demoralizing Racial Conditioning: The Experiment

History deposits many acts whether they are of humanitarian benefit or atrocity. None meets the historical example in evidence detailing subversive, oppressive assault as the 'perpetual, on-going' psychological assault on Afro-Americans[3]. Before the inception of America as a nation

[3] 'Experiment' assigned Afro-America with the affirmed support of the academia, intellectual global community.

the introduction of the Black race incurred the brunt of bondage, degradation. Slavery posited the shores of America's Jamestown Virginia in 1619. Forced to will another's purpose spoke of emasculation, defeatist attitude to resist reducing 'definitively' person to a state of subjugation; psychologically speaking. Nearly four-hundred years have witnessed the revision and continual implementation of the demoralizing psychological assault of the African race from the Wiley Lynch mantra and Vaudeville to the political [supplemental] legislation of the Civil Rights Act and Affirmative Action. Bold indictment of the supremacist scourge racism, evidencing the Black plight has been a perpetual denigrating psychological war experiment manifest of racial insecurity, bias to eviscerate mind and spirit.

Inferiority Superiority Programming

It is necessary to omit numerous atrocities for the book would extend into a voluminous catalogue of detailed occurrences. Although, residing in the halls of academia a trust of speculation supporting the theory of Darwinism persuades the catalytic evolution of man from ape. There are distinctions, but it has been evidenced this theory politicized the argument of racial inferiority referenced exclusively to the Black race. Subversive Darwinism demonstrates how the Black plight resonates of White insecurity in the attempt to demoralize 'the image' of the 'targeted' race (let us be clear) by assigning primordial status to a specific racial group, and to politically, scientifically propagandize racial identification as being less human; i.e. inferior to a proclaimed racial superiority. This auspicious claim, universally acknowledged of dubious intellectualism has in of itself subverted academia due to speculative, incredulous evidence.

Such proclamation were the driving force behind other published scathing opinion as Benjamin K. Hays 'Natural Selection and the Race Problem,' expressed, "the superiority of the American Negro to his African brother...is due to slavery...as the present attainment of the American Negro has been solely the result of his close personal contact with the white man...the negro has been domesticated... has never been a competitor but has always been subservient to the white race." Hypothesis propagandize and reinforce the stereotypical perceptions in Afro-America of inferiority in comparison to Caucasian.

Centuries of this type of demoralizing propagation have achieved caustic results. Psychological subversive technique has produced a 'cultural identity complex' throughout the global human scheme exclusively conferred upon 'people of color' to the point of consternation. It is particularly heard in rants of stereotypical loathing perceptions of one's own racial characterization. Observed in the treatment where outlandish displays of obdurate, vulgarity 'unprovoked' toward individuals 'unknown' of the same racial identification conferred an outpouring of consideration and respect upon *usually Caucasoid* demonstrating cultural complexity of self-hatred.

Dominant Subordinate Schematic

The case of demoralizing psychological self-hatred, inferiority finds its foundation in the domination-subordination syndrome. 'The Complexity of Identity' by Beverly Daniel Tatum provides a comprehensive assessment of the cultural identity complex. "Dominant groups...set the parameters within which [the] subordinate operate. Dominate group holds the power and authority...reflected in determining who gets the best jobs, whose *history* will be taught...whose relationship will be validated by society...the dominant group...influences...the structure of the society." This is a dictatorial to conformity, adapt to condition outside one's cultural base for acceptance. She further goes on to state, "relationship of the dominants to the subordinates is often...defective or substandard (social indifference can erode self-confidence, worth)...For example, Blacks...characterized as less intelligent than Whites...dominant group assign roles...that reflect the latter's devalued status...target group *internalize* [these] images...find it difficult to believe in their own ability." The psychological attack is meant to be both dehumanizing debasing where *subtle emasculation techniques* are affected throughout socio-economic, familial institutionalized frameworks dictating, practice of both the race and other ethnic classes, personally experienced, manifesting culturally. Traumatizing social influence aims at 'altering perception, behavior of the target through underhanded, deceptive, abusive tactics."[4] Jean Baker Miller, author of "Toward a New Psychology of Women" asserts, 'inequitable social relations...seen as a model for "normal human relationships." [it remains]...acceptable in any circle to denigrate a particular group, to exclude...from one's neighborhood or work setting or oppose initiatives that might change the power balance."[5]

Assessment speaks of deliberate deprivation, debilitating action of the authoritative, power structure to subscribe through various forms of social economic action disparaging, propagandize despicable imagery of undesirable targeted group for the sake of undermining the basic fundamental to exist; to live and prosper equitably within the cultural base. It is not enough to witness such an occurrence, but to experience within the social scheme.

Demonstratively although such assault on the mind has reinforced those victimized with resolve, resilience and as of such affirms what the adversary has set forth to do with bold intent. Hays admits what the war strategist

---

[4] Psychological Manipulation (excerpts) - Wikipedia

[5] The Complexity Of Identity: 'Who Am I' (excerpts)– Beverly Daniel Tatum www.whiteprivilegeconference.com

commits; "the black man…so long as he remains subservient his position is secure…as he [the black man] becomes a competitor his fate is sealed." This is the unmasking of the contrarian pronouncement of the state 'prosperity, equity, justice for all,' where competitiveness (mantra of capitalism) meted with, if necessary, fatal reprisal. He further goes on to say, "America requires a dominant and a subservient race. Should the black man become dominant, he would fall victim to "that vicious element which renders the home and highway unsafe for unprotected women." This relinquishes the posited claimed of controlling the mind uncontrollable and speaks directly to annihilation. Hays concludes with--"where the strong cannot control it will destroy."[6]

There are various socio-political, historical events that affirm the nefarious claim of those strong-willed, determined *to just exist*. The old proclamation, 'we come as friends' has never reigned more circumspect with ulterior motive than in modern day diplomacy. Demoralization is the traumatic force of perceptive mind manipulation which brings into doubt capability, purpose, identity eroding one's trust in self and fellow. Once the objective has been achieved it manifest into divisive elements of contention.

---

[6] Natural Selection and the Race Problem – Benjamin K. Hays Charlotte N.C. Charlotte Medical Journal, 1905

Psychology

Infect the mind, achieve psychosis, and stir the senses

The Iraqi war was a hostile takeover precipitate by the desire of a corporation to be multi-national in the global arena of business. "Characterizing Iraq as insurgency rather than 'civil war' highlighted sectarian tensions and fragmented control…removed the issue of breached domestic sovereignty from continued US troop involvement."[7] Iraq exists under 'perpetuated chaos-control mantra,' provides analytical assessment for Arab Spring initiative. This analysis maybe considered far-fetched but the results speak otherwise. Utilizing military forces for the sole purpose of murder, regional instability to justify a perpetual occupation, destroying assets, infrastructure i.e. to cultivate a military/business presence; gain access to natural resources, geo-political regional terrain for future strategic objective is irrefutable.

Observations hold the initial conflict was only the catalyst of a broader agenda as —— the conflict has fermented Afghanistan, Iran, Pakistan, into a war objective by geo-political proximity where instability confer on the whole region a reassessment of government, commerce, diplomacy. Military action causes evacuation, refugees fleeing conflict, insurgency of those joining the battle, exploiting the terrain; spillover is a predicted strategy of such to broaden the scope of regional conflict. It precipitates the *necessity* to involve neighboring states

---

[7] Development of Conflict in Arab Spring Libya and Syria: From Revolution to Civil War – Maya Bhardwaj pg.79

placing deliberate strain on diplomacy and agreement. This is considered a door to extend a campaign beyond the initial conflict.

This example does not detract from primary goal and objective but demonstrates the precipitous nature of imperative lending cause to future initiatives, hostilities. Dedicated planning at considerable length is affected to achieve the salvo of the first shot fired. This has been a virulent stage to affect debilitating blows, under mind cultural consensus in the pursuit of expansionist objectives. The disease has reached epidemic and mutated into yet another infection called the Arab Spring.

Objective

Demoralization speaks of 'discontinuity' rendering host incapable of functioning, organizing effort or ability to maintain harmonious conformity. Initiative's apparatus applied most readily in matters of state intrigue for the cause of fragmenting, weaken recognized body of administration. This is psychological war; subversive, psychologically, socially, economically debilitating.

For example, a nation state (government, administration) must be recognized in and of the people it presides over, acknowledged as sovereign by other nation states. Government presiding must be able to defend itself (civilian police, militarily) within without clearly established territorial borders, in effect protect its citizens. Territorial resources must be cultivated of the governing body for the sake of commerce in trade. Territorial infrastructure must be adequate to effect maximum output of commerce, commute, provide convenient conventional means of support (public services) for its citizens. These are the basic elements constituting the socio-economic political aspect of a recognized nation state.

Government, territory, military provide the impetus for prosperity and defense of adversaries set on undermining state. Intrusion and compromise occur by and through dissension, coup, external interference or all there mentioned. This is intrigue proffered of exploit, opportunity for the sake of overthrowing established order. The intrinsic element of government (for example) can be infiltrated and compromised of internal, external host actors, political, legislative law and or agreements, dissolution. Fragmentation, disruption of government administration renders its ability to manage the state impotent, provides avenues for corruption, stagnates, renders the function of administration inoperable. With lack of continuity, calls for transition escalate. Even where the various regimes (Libya, Tunisia, Algeria, Iraq, Egypt) exhibit stability, the power of the people (sans externalities) demonstrates 'sovereign control' over who shall govern.

Infrastructure is severely important as the Iraqi invasion demonstrates. "Iraq's power, communications, water, sewage treatment and health facilities have not been allowed to recover from the extensive damage inflicted.in the 1991 US led Gulf War. Years of stringent U.N. sanctions have severely limited reconstruction, leaving the country's services in such a state that they would disintegrate within days in the event of a US military strike, with serious consequences for the population."[8] Three (3) statistical finding standout in the Iraqi War conflict:

(2006 estimates)

1. Number of Iraqis who died of violence

   2003-2011: 150,000 to 400,000
2. Orphans in Iraq:

   4.5 million
3. Orphans living in the streets:

   600,000[9]

Intentionally directing the war strategy at civilian infrastructure, although a '*humanitarian crime*' by international standards exacerbated internal sect, factional tensions (discontent, insurrection) at the regime allowing for contentions which brought about various regime

---

[8] The Impact of a US War on Iraq's Civilian Infrastructure – World Socialist Web Site.com 11/04/02

[9] Iraq: The Human Cost – web.mit.edu

opposition, administrative, military defections as well devastation wrought on the most innocent of victims. Nevertheless, the strategy achieved objective most favorable outcome sort by U.S. and British forces. The assault has reap a psychological price on the diverse ethnicity domiciling in the region. Creating such internal strife in a sectarian vacuum as Iraq's poses dilemma for future growth and posterity. Internal ethnic strife caused by outside force will not tolerate those forces interference and will incur backlash where escalation in hostilities might lead to future conflict. Even thereafter conflict has cease and occupiers leave regime will likely reassert itself perpetuating the ideological conundrum undisposed.

Sociology

Establish dialogue, diplomacy, discourse divisive

Leaders weld power to the extent they are given control of human, natural resources to prosper, defend the state but it is a matter of influence that achieves a broadening of proposed objectives. An intangible emanating force exhibited by few in matters of state, evidenced by threat, or show of persuasive rhetorical force in matters of diplomacy. This is demonstrative of negotiations conventionally ripe for expedient resolve.

Of recent, conscience effort in resolution has been absconded therefrom for the sake of accusation. Intensifying regional tensions to increases internal chaos utilizing those discontent with the administrating governing body proffering rhetorical resolution, a weapon of instigation to outside adversaries.

Beholden of any initiative are the thoughts, aspiration, intentions of the actors involved in contention. The goals and objectives might conflict (where in some circumstances they agree wholeheartedly) to the extent it might be conducive of desired outcome but necessary to affect a personnel change. It has been evidence that unity, cohesion are an aberration in matters of state. It has presented itself as a tool of bartering as in 'you scratch my back and I'll scratch yours.' Patience and timing are paramount in strategizing.

Qaddafi (murdered), Mubarak (imprisoned), Musharraf (on-trial as of this writing), al-Assad (embattled) are a group of regional (Northern Africa, East Mediterranean) leaders (ironically neighbors) whom find themselves embroiled, being toppled from the perch of rule. It is an egregious action (Arab Spring) initiated by western powers in attempts to realign the geo-political landscape.

More than inclination has achieved this fetid coup. For example, Muammar Qaddafi, pursuits (assessed from his last speeches and pronouncements) were of economic concern. Qaddafi proposed monetary policy (oil for gold dinars) would have shifted predominate relevance of the dollar in various markets. Other than espousing a unification of common neighboring nations his goals were pragmatic and threatened very little in the global theatre. Commentators consider his aspirations the primer for his execution.

What precipitated such event must be scrutinized of premeditation. Strategic planning, initiated of an opportunity to strike. Fatal cause has been categorized to some extent. It seems that fatality is influenced of self-infliction or inflicted by kindred souls, in some cases both. Conduct resides in the annals of discontent and envy and plays on the offenders insecurities more so than the convictions of the victim. This is a weapon in the hands of puppet masters exploiting the ignorant, oppressed mind lending opportunity to affect the fatal blow.

Western powers have mastered the art of chaos management. Upheaval is a matter of calculated provocative inducement to affect a cause of exploitation and exhibition into crisis via propagated medium.

Rhetorical Nature of Diplomacy

Rhetoric, a persuasive technique spawn of psychology affects a conciliatory parlance in matters of détente. It is a diplomatic tool born of ancient to inspire persuasive discourse where negotiable parlance deem less negotiable. Rhetoric expresses much, delivers little, softens the critical blow, and may ease tension. Duplicitous nature requires all who embrace it take heed of its alluring aspects. It is psychological linguistic pretext in matters of discourse.

Rhetoric Critiqued, Analyzed

"Foreign policy '*discourse*' is thus active in the production of *common sense*, amplified and reiterated in the institutions and practices that it generates... Extending the olive branch expresses a desire to reach out for the purpose of establishing a social bond acknowledging things one of another have in 'common' for the sake of dialogue deemed critical in the pursuit of further discourse... Accepting such a definition of discourse privileges the rhetorical critic with a range of social practices and institutions that are open to the lens of discursive analysis..." As does policy and the writing of this book. Discourse lends itself to the critique' to effect response concerning matter of mutual consent. Provisional of terms agreed there upon conveyed cultural, humanitarian assistance, (as in NGO's funding) conferred sincere acceptance. Rhetoric in such matters conciliatory of true intent conferring a sense of commonality in the pursuit of goals yet realized leaving room for ambiguity, or amendment (*ex. Brownback Amendment @ pg. 41*).

...Discourse instructs a typology of motivation, and consequently, action...looks for the "terminal compulsion' that individuals (or governments) carry out – such as the creation of a policy that anticipates totalizing discourse and ironically contributes to consequences that such discourse imagines."[10] In matters of diplomacy rhetorical discourse is the showman' s game of opportunistic ploy in times of negotiations deflecting from the issues, provisional of true intent.

Expanding on the idea of democracy posits an excitation of intent proselytizing via imposition by and through rhetorical gesture. The rhetorical pronouncement, 'Saddam Hussein has weapons of mass destruction' persuaded the international community to participate, launch a precarious invasion wrought of an accusation. Bush administration acerbic claims exemplified the

[10] The Hague Journal of Diplomacy 2 (2007) 229-254 – "Arguing Public Diplomacy: The Role of Argument Formations in US Foreign Policy Rhetoric

devastating effect of rhetoric cadence. "Persuasion and judgment are two main values in the rhetorical democracy..." No matter the argument put forth, whether truth, falsehood (*allegation made by Sec. Gen. Libyan Human Rights @ pg. 42*) or spin it provides a bully-pulpit for the practitioner. With the stage set for contingencies of debate rhetorical expression has lower the drawbridge of the adversary to gain entry into the stronghold to reconcile difference, plunder or both.

Rhetorical Social Medium Outreach

Contemporaneous aspects of rhetorical democracy have evolve with the technology of their time eliminating conjecture. "...For example, in the beginning of modern democracy, political leaders might need to persuade the mass as well as now. However, they might not spend so much time or only defended during election campaign. *Some phenomena are wholly innovative.* For instance, the Internet, social media or online video have given today's democracy new possibilities or challenges. *This is qualitative shift.*[11]

This '*shift*' has become 'the global light in the darkness, inquisitor, demagogue of the millennial.' Imposing its influence on the masses of the disinterested; inviting them to the debate. Affecting a responsiveness that

---

[11] Academia.edu.com – A Preliminary to Rhetorical Democracy Chaun-wei Hu – pg. 4

would have gone unheard. "Rhetoric in media coverage *bi-directionally* interacts with the instrumental dimensions…is fed and changed by policy reports…publicizes and mobilizes …actions by…politicians, international NGO's…media figures to reframe and depoliticize…by semantically separating it…within the realm of international obligation."[12] Meaning rhetorical discourse, whether preferred or otherwise has become a prerequisite in utility allowing for the appeal of the message put forth. The 'quantitative shift' has been invoked as well. Rhetoric is no longer confined to pursuing cultural masses but the global populace, utilized to secure and acceptance or condemnation from a broader audience.

Status-quo; have-have not deprivation is a constant captivating issue, for example, in the rallying cry to rise up, rebel which leaders have no problem in responding to with show of force.

Intervention besets the primary goal. A crude pause-setting of events initiates the cause of probable imposition. Intelligence has become bold speculation in matters of international conflict conferring pretext to response and the action *sometimes* murderous. Fabrication of evidence detracts from the primal objective, [of recent] to affect criminal humanitarian acts (considered by

---

[12] Development of Conflict in Arab Spring Libya and Syria: From Revolution to Civil War – Maya Bhardwaj pg.79

international standard) directed at nationality or leader. Precipitating events requires exploitation of situations producing contention, strife of a sort with past occurrence of altercation yet to be resolved. Matters are broadcast outside there venue to solicit opinion for the sake of a broader moral assessment. Hence, what was an issue in locality has become of global inquiry.

The broader the assessment (from a global perspective) more likely an exacerbation of tensions will ensue. This provides for the instigation of actions (covert or otherwise) to increase attention on governing, ruling class for the sake of external involvement. In most cases the initial cause becomes mere happenstance now that leadership has come under question.

This is diplomacy proffered in such strategic initiatives as the Arab Spring. It is none the less age old psychological ploy demonstrating a potential as event is intensified for the sake of involvement. Elements of dynamic are proportionate of provocation, exploitation; provisional catalyst for escalation. Defensive posturing (rhetorical technique) requires the proliferation of internal forces to deter opportunistic aggressive response. While extended negotiations provide access to prospective parties of interest. Once a dialogue has commenced causation takes on a new dynamic. It becomes apparent that the nature of intervention is pretext to gaining access to intelligence, resource and defense capability. With the threat of military action (pronouncement of waging war) from an international community focus of contention is exacerbated to the point that logic dictates collaboration of an orchestrated assault within-without the cultural scheme has occurred.

Resolve, being far from the objective provides a cloak of concealment for the exacerbation of internal rivals set on irresolute age-old contention and strife allowing for the discontinuity of the cultural construct. Administrators, political actors are absorb in the complexities of their own ineffectiveness to broker or institute stability within existing governing body for lack of diplomatic compromise. Revealing ineptitude, weakness in ability to manage the state.

A grand example as of this writing (6-30-14), Iraq finds itself grappling with internal strife, the making of civil war. The prime minister elect (al-Maliki) in all things governing has no internal-external support in matters of state attributable to external geo-political diplomatic influence which dictated prime minister election and prescription of state management. This type of governance perpetuates the action of crisis and has instigated the birth of new factions (ISIS), growing insurgency where the objectives of controlling territory, resources paramount.

External forces (host nations) have affected a controlling hold, interest in everything Iraq.

Once again, allowing for a brokering of outside influences set to administer from a distant perch of observance. Action of such can never be considered favorable deriving from external forces where past contentions make for suspicion in crisis. Patience has its place in the matter of expedience dictated by such diplomacy, negotiations. This is the mind (fundamental) of war and the military strategist.

Destabilization is not a matter of causing strife to affect a momentary crisis, but a dynamic that allows forces internal-external to affect a perpetual controlling interest within the cultural mainframe. It is advantageous of the accusers to instigate cultural upheaval, ferment dissent for the sake of intrusion. The exaggeration of the situation are expression of rhetorical discursive medium, impromptu propaganda. Observation holds a diabolical aspect of causation exist within the very framework of each foreign actor internal geo-political construct in the form of various tariffs, agreements. Upheaval of such can affect debilitating blows, socially, economically to the overall cultural construct. Grecian austerity and its disruptive effects on the Euro-zone, repercussion felt globally make for such a case of total socio-economic reformation.

Multi-national corporations are key facilitators both economically and politically of Iraqi infrastructure having quasi-controlling interest in production of oil resources as well a de-facto political influence in the governing body. This type of compromise makes for dearth administrating leading to perpetual bouts of unrest.

Noble pursuit, ambition are beholden the age-old scheme lain before aspiring leaders dictating their approach. World leaders seem to forget they were tutored and conditioned by the very actors they depose relegating their political platform to 'different day, same old...' This is evidenced of the repetitive, reoccurring diatribe, contention and strife unresolved of time(s) past. It speaks of the mind-set posited of ancient and confirms the inert cognitive mental state. No body of leadership has demonstrated innovation to the contrary. This is perpetuation of the repetitive rhetoric bold scheme...guaranteed.

The 'same old' has led to inertia which from the cultural socio-political view perpetuates instability. This is not about progress, but profit and control. Managing chaos is a business intelligentsia considers their holy grail. Geopolitical strategist are global watchdogs searching and sniffing out possible outcome of past, present, future international socio-political economic occurrence to affect high yield on investment. In most cases diplomatic, militaristic in nature. Diplomacy allows for the dictate where the military provides the element of force. Consultation does not abscond from blood-letting, coup d'etat. Characterization of the geopolitical strategist is befitting of all in the arena of investment. Theirs is to anticipate the action, response, behavior of the have, have not dynamic whereby politic, initiative bring about catalytic result increasing strategist influence or take on profit or both. Diplomatic relation leads to the accessibility of controlling interest. Strategy proffers a need to those in time of crisis, and uncertainty. From this *consternation* comes gainful employment, profit (one form of another), fatality. Bottom-line for the strategist; end-game for those in which hostile action are directed.

In the geopolitical theatre outcome rivals that of a chess match. Observation, patience and the proper inducement can reduce a culture into perpetual cadence of internal confusion and conflict. It behooves the writer that outside aggressor's rhetoric has more influence and standing than the indigenous cultural leadership. This has made the geopolitical strategist de-facto global ambassador of leadership, whether president, prime minister, CEO, or philosopher of affairs international. These politico's ply their trade through unilateral tariff, agreement, course dialogue of compliance accusing nations of less stature with violation when they themselves instigate the atrocity by and through upheaval via internal asset, aggressive political economic action.

The primary weapon (of recent) spearheading these objectives purport to propagandizing the *terrorist platform* to a precipitous level where even casual criminality is posited the designation. The campaign is a rhetorical geo-political strategy in effect polling support for the indictment, execution of clandestine assault, murder of those directed. As of this writing, all extreme acts of aggression have been initiated by Western, NATO-powers. Justifying murder in of accusation without evidence. In this matter pre-emptive strike has conferred a sinister characterization on patriotism.

Shrewd the application maybe in the matter of war (recognizing some in the theatre are unaware or naïve they are at war) evidence of duplicitous action is overwhelming in diplomacy to the point of psychosis. It can only be said, "all is fair..."

Destabilization

There exist shards in the foundation, tattered threads in the fabric of the culture which force maybe applied as to rend asunder

Bnjhmn

Interactive nature of sociology confers on the recipient perceptive cadence influencing a directive course of action. Precepts (belief) manifest conferring a religiosity expounding a moral, ethical conviction demonstrative of conduct. It has become the harbinger of ancient cultural conflict persistent.

Cultural adhesion on a global scale is illusive but a prize asset of Western, NATO-bloc. The lack thereof provides accessibility, a door which allows outside interest may ply their assault. This mechanism provides the impetus for exploiting diverse interest of conflict occurring within Muslim nations destabilized. Animosity allows for instigation; the escalation of matters better address internally. War strategist will to exploit any internal conflict for the sake of intrusion. What has been demonstrated from a socio-cultural observance suggest a greater contempt for one's brethren than those exacerbating conflict from without.

Individual preference and belief defines humanities propensity for conflict, confrontation. The scholar, intellectual, academician, theologian alike have parlayed virulence to divine order in their perception (not absolution) of truth. Although humanity demonstrates a tolerance course action is reserved for matters ill-resolve. Even examples of American cultural conflict, from Civil War to Civil Rights to Civil Liberties propose they are continuous, ongoing; perpetual. They are of a similitude, espousing godly dictate introduced of men. Conferring on matters divisive have dissuade humanity to partake thereof what is right and proving for the sake of civility so as to invoke conjecture, opinion. It has become a weapon in the hands of one adversary. The weapon of thought; belief. To separate, rent asunder the cultural identity, even to blood. Explosives and bullets have no claim on the matter of *mind* for it is *here* they find place ill-resolve.

"There will be wars and rumors of wars," partaking of doctrine thorough observation of the tools of media posits a circumspection of authenticity, creditableness. The onset of military operation in Iraq, evidence to follow made intelligentsia, media desk suspect, lacking all creditability. Removing the hype allows for a better analogy of events precipitated.

Sectarianism

"...is bigotry, discrimination or hatred arising from attaching importance to perceived differences between subdivisions within a group, such as between different denominations of a religion, class, regional or factions of a political movement...At other times, sectarianism may be the expression of a group's nationalistic or cultural ambitions, exploited by demagogues."

Wikipedia.org.com – Definition: Sectarianism

-Essay is void of religious preference and opinion-

The imperative is derivative of politic, a science of diverse opinion divest of all things absolute. The division bell exacts the price. Western powers are quite pleased with the results. A conflict fashioned pitting cultural rivals one against the other for the sake of influence. This is war none the less, appropriated...

Definitive element posited portrays the matter to an exactness defying belief. It behooves one that foreknowledge is absent the matter spawn of schism. A spectacle void of cohesion, unity fermented and violently exacerbated by calls of divine cultural identification.

...orchestrated, strategically designed in an of demagoguery

with the assistance of the host adversary.

History

Arab Spring initiatives are ripe of ancient discourse which began in the year 632 following the death of the Prophet Mohamed. The Prophets campaign on pagan tribes in the last ten years of his life united all of Arabia. Succession of rule would pass to the leader's son, (his was a spiritual as well politic calling) however Prophet Mohamed had no son, only a daughter. Prophet Mohamed's adherents thought his closest associate, Abu Bakr, should preside over the nation. They were Sunni and the majority at that time. The minority 'Shia' sect thought the Prophet's closet relative, his son-in-law and nephew Ali, should succeed. This debate spawn the division as response to the debate turned violent as Mohamed's widow Aisha (daughter of Abu Bakr) lead troops against Ali whom was killed, as was his son Hussein. These historical events have been the life blood of the centuries-old conflict, schism being affected as a political weapon of divisiveness.

Survey of Cultural Sectarian Difference

Prior assessments allow for strategic planning and orchestration of destabilizing actions and continues for the purpose glimpsing reaction of the internal host cultural identity.

"Muslims in the Middle East and North Africa tend to be most keenly aware of the distinction between the two main branches of Islam Sunni and Shia...countries surveyed in the region, at least 40% of Sunnis do not accept Shias as fellow Muslims...do not believe that some practices common among Shias...are acceptable as part of Islamic tradition. Only in Lebanon and Iraq – do large majorities of Sunnis recognize Shias as fellow Muslims and accept their distinctive practices. Outside of the Middle East and North Africa, the distinction between Sunni and Shia appears to be of lesser consequence. Central Asia, Southern, Eastern Europe as well Indonesia, which has the world's largest Muslim population, 26% describe themselves as Sunnis, compared with 56% who say they are 'just a Muslim.'"[13]

Sectarianism is a social schematic of divisive posturing where course variations institutionalized within core foundation are of fixed irresolute conjecture. Rivalry can be observed within a racial cultural mix, for example 'white versus black,' political affiliations (democrat, republican), governing philosophies, ideologies (communism, socialism as oppose to fascism, capitalism) having similar polarizing effect. The common core element of human existence proffers the only relevant resolve in such matters as we observe and conclude that, 'we need to

---

[13]Pew Research – Religion & Public Life Project: Aug. 9 2012 The World's Muslims: Unity and Diversity pg. 4

eat, seek shelter from the elements, sleep and eventually die. A tolerance is maintained in matters uncompromising.

Arab Spring initiative rings analogous to the story of the 'three little pigs' (as not to invoke a semblance of ethnic identity, culture or otherwise) as the 'wolf' stands at the door knocking. Security depends on the strength of the foundation (house) whereas internal factors of administration require bold decisiveness to prevent 'the beast' from entering, inflicting deliberate, fatal harm. Of evidence, Northern African nation states have succumb to intrusion due to weak foundation (house of straw), infiltration (poor administration). Broad extensive nature of initiatives are testing the stability of other neighboring nations such as Syria whose endurance (house of wood) demonstrates a parlance of resistance to the intrusion and assault. Whereas Iran displays an imperviousness (house of brick) that speaks of soundness of foundation, administrative competence. What has been witnessed is external forces invoke catastrophic event to affect favorable outcome in matters of transition exploiting civilian populations to justify unconscionable humanitarian intervention.

Timeline

Algeria, Tunisia, Libya, Egypt. The instigation, assault of unrest has been conferred an identity none disputes. From December of 2010 to March of 2011 (ninety-day period). All of Northern Africa has experienced transition in government Fig.1 (mostly hostile in nature) of the Arab Spring initiative.

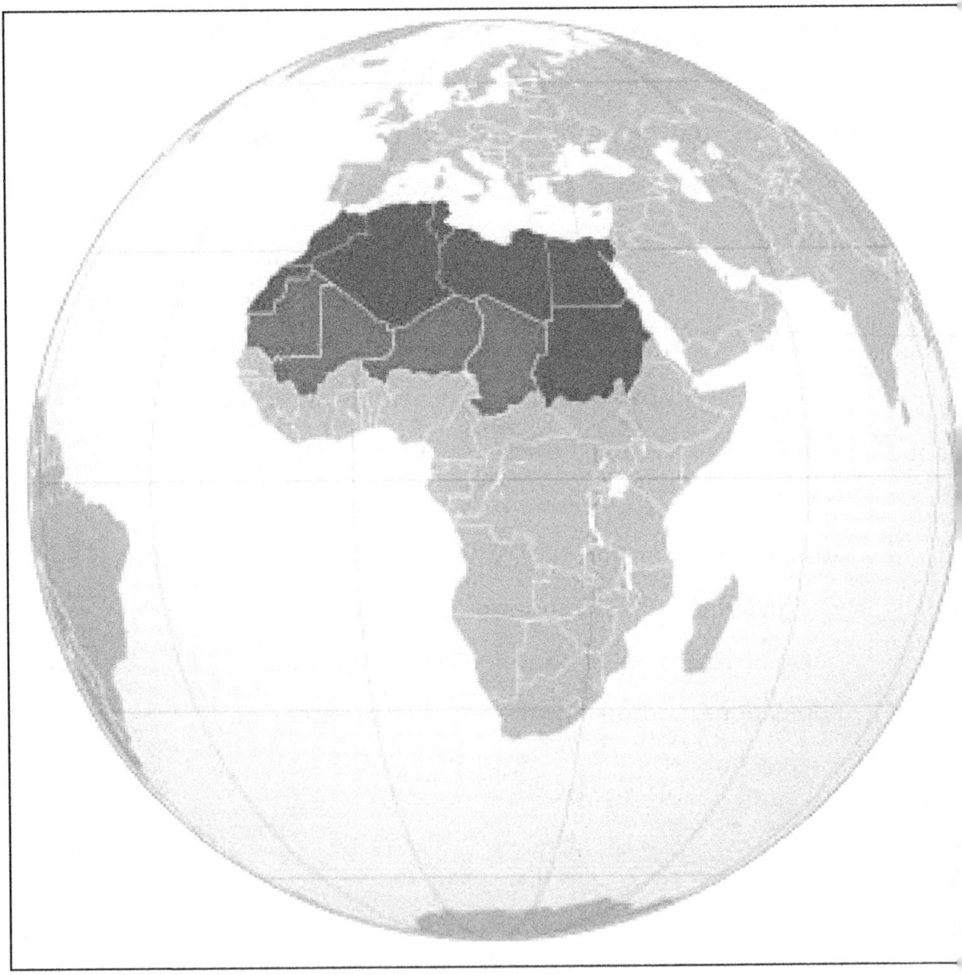

Northern Africa Nations Affected by Arab Spring Initiative

Causation; Instigation

A primal factor sparking the unrest in Egypt, Algeria, Tunisia, Libya, and Syria has been the influence of NGO's (Non-Government Organizations). The organizations have come under prosecution, intense scrutiny in Egypt by the Mubarak, Morsi governments alike for their accused involvement of funding street protest. It has been evidenced that every occurrence spearheading regime change in the proceeding list of countries were instigated of planned, orchestrated protest.

Charges were brought against four U.S. NGO's in December 2011. Even with the fall of Mubarak, Morsi governments and the election of Abdel Fatah al-Sisi former commander of Egyptian armed forces the organizations have come under scrutiny. Drafted law and release thereof in late June 2014 by Egypt's Ministry of Social Solidarity would allow the government to dissolve NGO's that "threaten national unity." Cause of unrest has been realized by those past and present, inside and outside of

Arab Spring Timeline

December 2010 – December 2011

Mohammed Bouazizi set himself on fire; protest ensue

Tunisia	President Ben Ali goes into exile	RCD Party dissolved	Ben Ali sentenced 35 yrs.	Tunisian hold first election	
Algeria	Major protest; 2 killed				
Libya	Protest begin in Benghazi	NATO commence operations	Rebels advance; take Tripoli	Qaddafi killed NATO cease ops	
Egypt	Mass protest begin	Mubarak steps down	Mubarak goes on trial		
Syria	Mass protest begin	Assad rejects Arab plan refuses to step down			
Bahrain	Protest begin; Suppressed				
Jordan	Tunisian protest encourage Jordanians to protest				
Yemen	Mass protest Begin	President Saleh injured; leaves country	President Saleh resigns		

Dec Jan Mar Apr May Jun Jul Aug Sep Oct Nov Dec

Fig. 1

Administration and Congressional legislative action provides the impetus. "The Egyptian government, however, administration denounce the United States providing this assistance directly to NGO's, noting its view that USAID was violating the terms of a process that the two governments had outlined in 2004. Although in 2005, Congress approved amendment to the Consolidated Appropriations Act (the "Brownback Amendment")[14], which stated that for "U.S assistance for democracy and governance activities in Egypt, the organizations implementing such assistance and the specific nature of the assistance shall not be subject to prior approval of the Government of Egypt."[15]

Bilateral agreement of 2004 would seem somewhat a pretext (the opening of the door) as to affect infiltration of the cultural host for ulterior motive. Breach lies in the fact the other party to the agreement were uninformed of the amended act. This is not a misstep but well calculated approach to achieving the objective. Excerpt of Black's Law dictionary states, 'a bilateral agreement can be "positive" or "negative" - you "will" or "will not do something." It further goes on to say, 'both parties will suffer if the agreement is not fulfilled. Funding protest by and through

---

[14] H.R. 4818 pg. 168

[15] GAO Testimony – Before the Subcommittee on the Middle East and North Africa Committee on Foreign Affairs, HR DEMOCRACY ASSISTANCE Lessons Learned from Egypt Should Inform Future U.S. Plans. HR Hearing 'Struggle for Civil Society in Egypt. 24July14 pg.11

NGO's having foreknowledge of the cultural religious divisiveness only confers a course strategic posturing with intention to cause strife. Diplomatic political rhetoric has rifled the cause for upheaval castigating all directed in its path. A prime *instigation* of the Libyan conflict assailed a petition signed by 70 NGO's spearheaded by the former Secretary General of the Libyan League of Human Rights demanding that the United States, European Union and the United Nations 'take immediate action to halt the atrocities being perpetrated by the Libyan government against its own people. The petition signed in earnest by the supporting NGO's invoked the "Responsibility to Protect" doctrine conferring on the council the 'act to respond' to 'humanitarian crisis.' *Less inquiry of accusation* by UN Security Council precipitated actions leading the overthrow and death of Col. Qaddafi. Although the Secretary General of the Libyan League of Human Rights *accused* the Libyan government of the atrocities the Secretary General's testimony and the petition *less investigation* caused the suspension of Libya from UN Human Rights Council where the matter was taken-up by the UN Security Council for military intervention. "A young filmmaker interviewed the Secretary General of the Libyan League of Human Rights referencing how he gained support for his petition. He replied, 'all NGO's are acquainted.' They are 'government funded and sponsored,' meaning their actions are influenced by the government funding them. Ask months later about his 'claim of atrocities' evidence thereto, his response was, 'there is no evidence!"[16] These are not matters of necessity but desire utilizing deceptive cadence, cultural strife less important. Clearly, this demonstrates why it is unnecessary to dissect religious elements (a *resource* of contingency) where cause pertains to diverging opinions, political expedience. The objective is to utilize diverse cultural element for the sake of achieving geo-political (psycho-socio economic) goals. Globalism, the political objective, goal speaks of the dominance, control of geographic terrain, resource by the elite superpowers (whom are in concert, agreement) and their multinational conglomerates. Once the global objective has been achieved maturation will purport to religiosity of autocratic materialism. A scheme merely to cultivate the whole of natural resource which includes humanity.

---

[16] Lies Behind the "Humanitarian War" in Libya: There Is No Evidence! – Oct. 19, 2011

Syria: Proximity; Cause

Scenario: pedestrian and the speeding car. You, the pedestrian observe the vehicle speeding toward your person and are aware that time is of the essence. You of course get out of the way. Thus the law of physics dictate action in matters of kinetic energy. Location as well politics, however is fixed as one's national, indigenous geographical origin. Arab Spring initiatives demonstrate a coordinated, scripted assault of the region directing course objectives at regimes considered a hindrance in the agenda of global realignment. It has not been difficult of Western-NATO bloc to affect the call to revolt observing internal governance placated regional leaders with rhetorical assertions allowing for lackadaisical intelligence, security countermeasure.

Regional leaders are quite aware of cultural contentions of diverse nature knowing well they have and could escalate into civilian-wars. What makes Syria different is a Sunni majority is ruled by a Shia minority (Alawites) sect which President al-Assad, his army officer elite belong thereto. Syria's population is a culmination of "various religions, sects, and ethnic groups: Arabs, Assyrians, Kurds, Armenians, Turkmen, Circassia's, Muslims, and Christians, Sunnis, Shiites, Druze's, and so forth. These groups had been held together by Hafez al-Assad and later his son Bashar. If the son's regime were to collapse…this might bring about the fragmentation of the country, resulting in extensive massacres."[17] Resilience could well be subscribed to the internal indigenous population as a reinforcement of al-Assad resistance to insurgency, revolutionary forces. However, the divisiveness of religious sect still affect an uncertainty on the destiny of the troubled region with Jihadist al-Qaida on the Sunni side and Hezbollah militants on the side of the Shia.

Nonetheless, sectarianism has become the penultimate weapon of choice. Decisive measure allows for external humanitarian intervention provisional of international law 'responsibility to protect' (R2P) whereby sovereignty of a state is [not absolute right]; forfeited when a sovereign state fails to protect its people. International humanitarian law has become the edict of 'undermine/overthrow' in matters of regime change. It is a lesson the Syrian government and people are submerged therein review of.

---

[17] 'The Vicious Schism Between Sunni and Shia' www.independent.com.uk

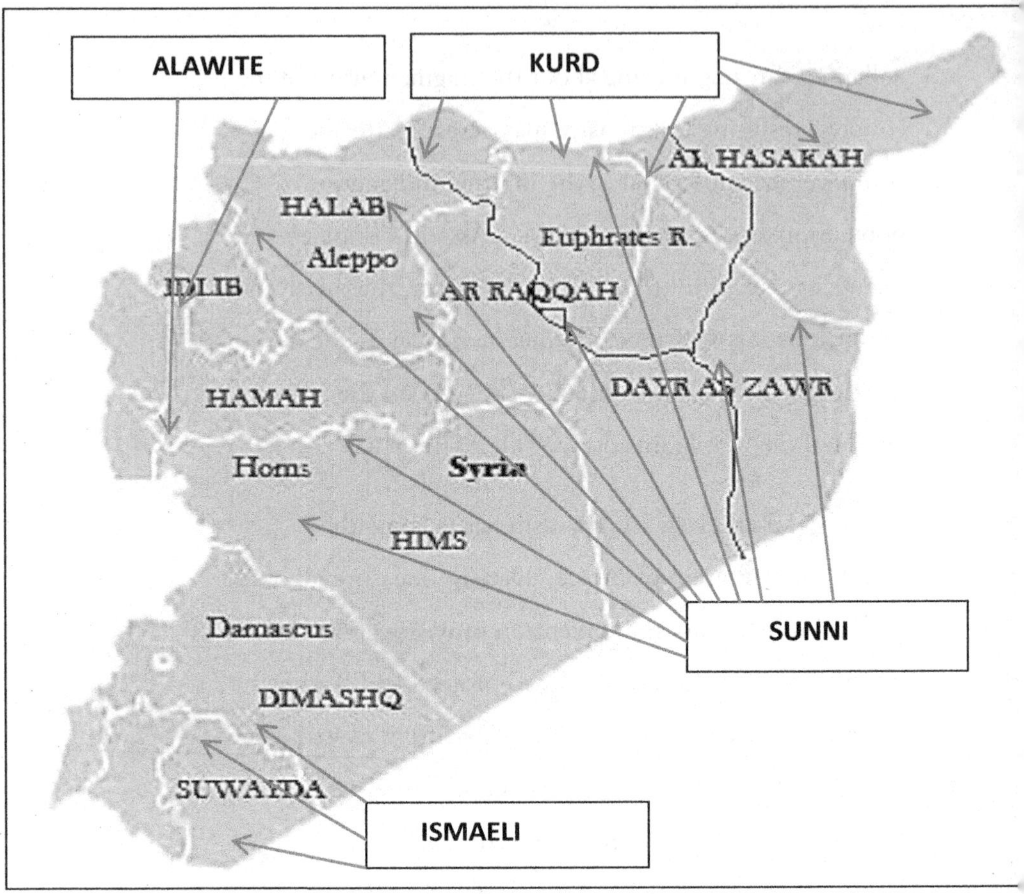

Syrian region dominate ethno-religious sects.

Excerpts from the Congressional Record 113th Congress

U.S Arms Syrian rebels -- (HR – June 19,2013)

"Mr. Speaker, there is a war going on in Syria. Some call it a civil war...ruled by the dictator Assad...Several rebel groups, and we're still not sure who all these people are, are trying to remove him from power...You have the Syrian Government supported by Iran and Russia...You've got Saudi Arabia; Qatar; you've got the

Muslim Brotherhood from Egypt supporting the rebels...The President has said we're going to give arms to the Syrian rebels."

Here's what the President said: "We're not taking sides in the religious war between Shia and Sunni. Really, what we are trying to do is take sides against extremist of all sorts."

Here lies the pronouncement of the politic. Of an interpretive: 'Due to a continual cultural religious divisiveness opportunity allows for regional strategic adjustment in the matter.' Although cultural ethnic disparities exist (have existed for decades) Syria finds itself in direct line of an expansionist campaign where initiatives require old-world regimes be absolved in and of future pursuits. "America established the set of alliances between countries of the Middle East. The White House achieved its goal, which led to conflicts between Muslim and Arabic countries."[18] Propaganda has led the way of revolts of the Arab Spring but many of the old-regime outside the conflict behind the scene are orchestrating, perpetuating the revolution(s) for the sake of influence. One nation has been conferred a catalyst in the ongoing initiative. Qatar evidently is a primary source of funding to the Syrian revolution. "A perception is taking root among a growing numbers of Syrians that Qatar is using financial muscle to develop networks of loyalty among rebels and (to) set the stage for influence in a post-Assad era." The article goes on to say: "Syria is also the culmination of an opportunistic

---

[18] Academia Education – 'Geopolitical Expansion of the USA'

foreign policy which saw Doha (seat of Qatar government) become the unlikely backer of other Arab revolts in north Africa – and a friend of those who emerge as winners..."[19]

Continuing Excerpts...

"...What is the goal here, to put another rebel group in power in another country? You know, we've kind of forgotten what we did in Libya...We support the rebels who overthrow the Libyan President...We don't know what has happened to those weapons that the United States gave to those rebels."

Libya

Valuation lies in fact internal contention and strife will achieve the objective. Weaponry is secondary to what fuels and precipitate action. Science of 'divide and conquer' portends to nothing more than disagreement between host rivals. Objectives of the external see opportunity in exacerbating the contentious element of internal rivalry. Beholding to this is that external forces both local and abroad are coordinating efforts in such manner as to be persona-non-gratis. Former governments (given the fact streams of medium existed of the initiatives prior thereto) seem oblivious to the events leading to their downfall lending to a perception that intelligence was incompetent, lacking or complicit of the occurrence.

To Western- NATO forces sectarianism purports to that of a business proposition, the like of a new cash-crop

---

[19] Financial Times – "How Qatar Seized Control of the Syrian Revolution" May 17, 2013

discovery in need of proper cultivation which requires plowing the field; a sowing may commence.

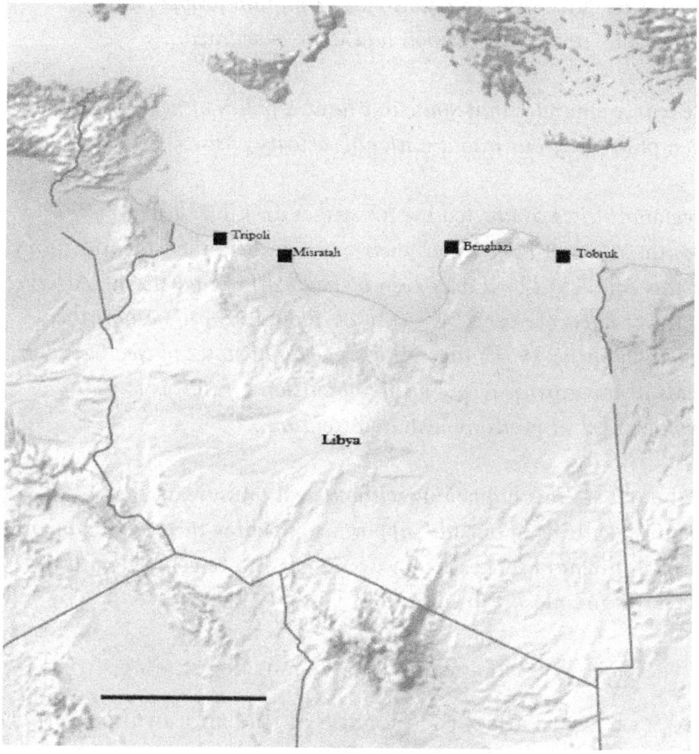

Libya's seat of Government Administration

Excerpts from the Congressional Record 113th Congress

Hearing: Committee of Foreign Affairs – (HR 9-10-14)

Libya's Descent

"3 years ago...a successful example of multilateral engagement– NATO and our allies all working together to stop the slaughter of civilians...free an oppressed people from dictatorship...chart a prosperous path forward for this country..."

"Unfortunately...Libya has become chaotic, violent, awash in terrorist organizations...militias...borders allow for easy transit of people, of weapons, of money,,, across North Africa to Gaza to Syria to Iraq...you have rival governments now in Tobruk and Tripoli making competing claims of legitimacy...the last four months we have seen an additional ¼ million people flee Libya...our embassy in Tripoli had to be evacuated..."

"It was a reminder that you can't have a policy of neglect that then plays itself out into a national security crisis..."

"regional states conducted the air strikes on Libya last month...attempting to affect their interest...help their favorite proxy in this conflict...Libya may even be headed for a partition...Algeria or Egypt may intervene...We cannot allow Libya to become the Lebanon of the 1970s and 1980s...regional states played out their feuds at the expense of the local population...if that is to be avoided...Libya needs immediate attention..."

"...I don't see this happening without real pressure on the factions...cutting of outside support...compelling their disarmament through threat of force...action would have to be coupled with programs to unlock Libya's wealth."

As does the farmer, the war strategist affects the proper tools to affect the plowing of the land, in this matter NATO and the rivalry of sectarianism.

In the case of Libya, Col. Qaddafi proposed legislative acts (nationalization of regional oil, currency reformation, wealth distribution) would have set the stage for bold transformation of regional, global economies to the point of consternation. Bold disruption of the status-quo affects harsh response as his assassination (descriptively) narrates. Published accounts allege an agent of the French secret service delivered the fatal gun-shot to the head of Qaddafi during his capture by none other than National Transitional Council (recognized by Western-NATO forces as the legitimate governing body) and rebel regional forces. The execution-murder, alleged to have been afflicted by democratic nationals (referencing the Secretary General of the Libyan League of Human Rights) affirms a barbarous hypocrisy looming over the global elite.

Col. Qaddafi proposed legislation spoke exactly to the congressman proposal; '*unlocking Libya's wealth*'. Libya is well known as the oil magnet of the Mediterranean. Although its oil production (530K as of 31Aug. 14)[20] down to 1/3 post Qaddafi era, it natural resource wealth takes second stage to it geographical location. Map above details the administrative localities all firmly situated on its coastal region allowing for unlimited commercial activity, prospect for becoming an epicenter of trade in the Mediterranean, prospective considerable probabilities.

War strategist seek the ultimate outcome in achieving the objective; total defeat or surrender, hence autonomy. It would not be conducive to provide a support mechanism where such cunning has conferred upon the opponent a supreme weapon (one of internal conflict). Various cultural elements were exploited in the matter as such factions existed within the cultural mainframe.

---

[20] Libya's Crude Oil Production – ycharts.com Aug. 2014

These dynamics allow for causality:

"1) *The nature and the action of the previous regime,* including the rise to power, makeup of the regime, public access to information and political dissent, and amenability to reform;

2) *The territoriality of conflict,* or the division of the state into government, and rebel controlled territories with loosely or clearly defined battlegrounds, and the ability of the players to gain local support and resource access in these home bases;

3) *Militarization of government and rebel forces,* including the loss of monopoly of force by the government and the ability and willingness of the rebel forces to arm and train its members through either internal or external means into cohesive and united troops;

4) *International influence,* in the form of state, international organization, and 'NGO approval,' aid, and/or intervention, and the nature of international media coverage influencing policymaking; and

5) *Regional players,* including the approval of bordering nations, regional organizations, and the flow of goods and services through borders."[21]

---

[21] Development of Conflict in Arab Spring Libya and Syria: From Revolution to Civil War – Maya Bhardwaj

"National Transitional Council formed in Benghazi in February 2011. Qaddafi's justice minister defected, became chief of the NTC 'quickly developed strong ties with Western, Arab alliance...France recognized the NTC with an amazing speed...when Qaddafi was officially declared dead council was declared *by outsiders* the only legitimate representative of the Libyan people, without a referendum, without consultation of the Libyan people despite warnings of infiltration by radical rebels. Shortly after installation, NTC was blamed for being 'helpless and weak' against the (popularized) militias...urged the people to respect the law they do not have in the newly liberated state, infested with weapons and militias, infiltrated with sponsored radicals, plighted with cross-border *contraband,* and inflicted with chaos and tribal division."[22]

It practically removes the adversary from the field of battle and assures confrontation, insurrection. Arab Spring initiatives are phase-stage pronouncement orchestrated of external forces infiltrating the cultural mainframe. Events are penultimate of the formulized strategy to under mind the governance of the Northern African region in totality provisional of geopolitical expansionism.

---

[22] Temehu.com – Interim Transitional National Council 5/3/11 – 8/8/12

Aftermath

Four years of internal conflict has affirmed the pretext espousing call for regime change. As of 2015, Arab Spring initiative have become the quagmire of regional irresolute instability the former vice president[23] described reference to invading Iraq. Fueled by Western-NATO alliance and their regional conscripts. Exacerbating the conflict has posited an even more pernicious culture of extreme practice.

ISIS

They have advanced through the region unscathed. Display a heinous desire to execute exclusively ethnic rivals affecting an alliance (delinquent) of nation state leaders to confront the proliferation of its influence. Their motions are uncanny, fluid occupying terrain as though they resided therein. They have advanced through the region with a tactical precision. They are perceived (by some) of an external alliances, (asset) as their campaign goes unchallenged. The denouncement of characterization (subtitle) denotes an ignorance to those subscribing identity (provisionally of suppose relationship) thereto cause. Their pronouncement of identity has been conferred and sanctioned. THEY ARE ISIS.

---

[23] Cheney 94: 'Invading Baghdad Would Create Quagmire' Youtube: Uploaded August 10, 2007

Timeline Narrative

Discontent and proper motivation can affect intended result. Iraq provided the perfect incubator, platform during the capture in 2003 of Saddam Hussein emerged the affirmable Abu Musab al-Zarqawi and al Qaeda. It was alleged his campaign was to promote sectarian war with the majority Shia community during his short tenure (2004-06) prior to being killed, martyred in American-lead airstrike. Abu Ayyub al-Masri succeeded al-Zarqawi and established the Islamic State conferring leadership to Abu Bakr al-Baghdadi.

Of a mentioning, after the execution of Saddam Hussein in 2006, Bush administration initiated strategy (early 2007) to increase the number of troops in Iraq by 20,000 (called 'the surge') although considered a success many critics attributed such to the fact factions and insurgence began fighting amongst themselves for power allowing for a further polarization among religious factions.

Islamic state suffers loss of Abu Bakr Al-Baghdadi and Abu Ayyub al-Masri, killed in American operation in Iraq April 2010. ISI merger (2013) with Syrian militant group Jabhat al-Nusra reject by its leader Abu Mohammed Jawlani whereupon infighting between al-Nusra and ISIS brought renouncement of ties (2014) from al-Qaeda.

ISIS capture, occupation during the month of June 2014

ISIS launched campaign throughout Iraq in June of 2014 gaining control of key northern cities, especially Mosul, capturing communication networks, airport and the governor office. ISIS initiative allowed for the freeing of prisoner advancing their numbers through conscription an objective put into effect with the May 2014 kidnapping of Kurdish youth for the purpose of indoctrination into the teaching of radical Islamic theology.

Withstanding Western-NATO alliances in Iraq, ISIS declares the caliphate (which by edict) erases all state border of Muslim states conferring al-Baghdadi (self-declared) leader in authority over its population.

As events occur, the actions of both Western-NATO alliance and ISIS become more conducive, coordinated toward the excitation of events outside the borders of Iraq, namely Syria. As the U.S. increases its military present by some 300 troops (including intelligence advisers) essential for securing U.S. Embassy and airport in Baghdad the proximity of ISIS succession and its uncontested expansion (insurgency) into the Syrian conflict ring suspect of conscription by and through agreement.

Early July of 2014 witnessed the expansion, successful occupation of the Islamic state (ISIS) directly into Syria via insurgency. Not less than two weeks had eclipsed and on 3 July, ISIS controlled major oil field in al-Omar Syria, had captured cities between Deir Ezzor and the Iraqi border. Their advancement claims stoked a number of killings and the capturing of another oil field 17 July, in Syria's Homs Province. This was conquest and 'public relations' befitting the ancient greats.

ISIS insurgency, seizure of resources in Syria July 2014

ISIS, media blitz provide perfect backdrop for diabolical expression due to war fatal consequences. ISIS media campaign, (thoroughly promoted by Western media) provided the visual effects of brutal executions, such as the 19 August decapitation of U.S. journalist James Foley, missing in Syria since 2012. Video Beheading, Steven Sotloff, 2 September. September 13 execution of British aid worker David Haines. Video Beheading, British captive Alan Henning, October 3. Execution of 322 members of Albu Minr tribe by ISIS in Iraq, November 3.

Whereupon CIA report released 11 September 2014 announced "the number of people fighting [for] ISIS more than three times the previous estimate." Taking full stock of it 'heinous popularity,' ISIS leader Abu al-Baghdadi is alleged to utilizes social media posting audio message November 13, expressing 'U.S. led-coalition is "terrified, weak and powerless.'

ISIS continues its brutal campaign, beheads American hostage peter Kassig, November 16. ISIS demand 200 million dollar ransom, 20 January 2015, from Japan in exchange for Japanese hostage, Kenji Got and Haruna Yakawu. January 24, online video shows hostage Kenji Got holding a photo of beheaded hostage Haruna Yukawa; demand made for the release of terror suspect Sajida al-Rishawi in exchange for Kenji Goto. Captured Jordanian pilot, Moath al-Kasasbeh, shown of video February 3, burned alive while confined in cage. In retaliation for the murder of Moath al-Kasasbeh, Jordanians carry out airstrikes, February 5, over Raqqa, Syria bombing ISIS facilities.

U.S. President Barack Obama petition Congress, February 11, to formally authorize use of military force against ISIS.

Video of the beheading, February 15, of 21Egyptian Christians by ISIS on Libyan Beach released. Egyptians stage airstrikes February 16, on ISIS targets in Libya in retaliation. ISIS Libyan affiliate Wilayat al-Barqa claims responsibility for three suicide car bomb blast, February 20, killing 30 in Gobba, Libya.

ISIS utilized its popularity on the internet to spearhead its propaganda campaign with the filming of ground fighting, captures, execution knowing the global populace thirst for such brutality. ISIS release of video February 22, of Peshmerga fighters in cages paraded through Iraqi street delivered not just a local message but was resonated to its global audience as the man threatens, "We say to you Peshmerga, leave your jobs or your fate will be like these, either the cage or under the ground."

Revelation of the identity of the executioner (whom spoke with British accent) of the Western hostage(s) took on a media blitz when it was discovered that in fact he was Kuwaiti-born Londoner Mohammed Emwazi aka "Jihadi John."

ISIS campaign seized on the speculation of the opposition confirming a dubious nature of sorts, which allowed for the expansion, underpinning, strategies which led to embarrassing dysfunctional foreign politics.

March 1, 2015 brought the release of 19 prisoners of Christian faith. On March 4, a man was thrown off a building in Syria for allegedly being homosexual conferring preference dictate cause of action.

Desecration of antiquities subscribed a loathing of identity to the very act of the bulldozing ancient city of Nimrud, March 5, to mention the bombing destruction of Jonah's Tomb a holy site in Mosul.

There can be no doubt a great deal of planning, negotiation forthrightly occurred prior to the campaign of the Arab Spring. Studying the timeline of occurrences the actors, foreign powers, resources and mechanism involved, affected, what was/has been accomplished by what most consider a religious rogue militia infers otherwise. From Iraq, through Syria, Libya, having a bold influence in the whole of the Middle East, Northern Africa asserts the force from within conferring without.

Nigeria's Boko Harum has caught the fever of the cause and its leader Abubakar Shekau has pledge allegiance March 7, to ISIS. March 12, 2015, an audio message released by ISIS proclaimed the caliphate has expanded to western Africa and ISIS leader Abu Bakr al-Baghdadi has accepted Boko Harum's pledge of allegiance. ISIS continues its assault bombing Iraqi army headquarters, the same day near Ramadi, killing 40 soldiers.

Span of twenty-four months propel a terrorist cell descriptively into one of the most effective rebel groups in the history of armed conflicts. It is not to go unmentioned that various elements (Western-NATO alliance, opposing factions) by and through 'political course concession' provided and allowed for the ascendency of ISIS as supplement to Arab Spring initiatives, thus for the perpetuating sectarian hostilities.

Notwithstanding claims, ISIS has been quite effective in its campaign. It has all but initiated a broad assault in and of the Arab Spring initiative. Islamic state has affected a collaborative effort where its appeal has drawn to its cause regional factions willing to align themselves extending the Islamic state influence and reach throughout the region and beyond via proxy.

Nation state leaders within the region must confess their ineptitude conferring on such matters where examination of reports of beheading 21Egyptians by ISIS in Libya has brought to bear the necessity to confer and affect a joint resolution in combating such atrocity. Speaking by phone to the Egyptian TV channel 'Dream,' Major General Commander of Operation Dignity Libya said, "Libyans are helping the Egyptian army take revenge on these terrorist forces because Cairo is considered to be a Libyan city."[24] 'Collaborative effort' (a foregone conclusion until recent) speaks of probable resolve where proper response relies upon conventional weaponry necessary to achieve objectives. This can hardly be the case where permission to arm oneself is required. Libyan UN envoy, Ibrahim Eddabbashi has appealed to the UN Security Council to lift arms embargo stating, "Because the Libyan army is launching war against terrorist organizations in Libya, some

---

[24] Libyan Army Ready to Support Egyptian Military Intervention Against ISIS – The Tripoli Post 2/16/15

of them have pledge allegiance to ISIS, it is in the interest of peace and security in Libya and the world to simplify the procedures for the army to obtain weapons by either lifting sanctions or by facilitating procedures for exemption from the ban."[25] Resolution of such nature speaks of uncertainty purporting to objective clarity. The opponent and the people observe closely the demure, indecisive nature of military. Islamic state as any other assess such negotiation in elation for it represents to the people a weakness, ineffectiveness of the leadership. Notwithstanding the fact foreign actors of the international community are the orchestrators of the impending regional conflict.

ISIS accomplishments are being observed of the people in reference to the perplexed nature of the ruling class; leadership. Decisive issue resides with the people provisional of an acceptance, acknowledgement of what is deemed best for the whole of the cultural base. Sans strategic planning, (both political and militaristic) what is being demonstrated in the theatre of battle by ISIS confers a confidence of diplomacy to the cultural masses as various factions come together to fight and defend against what the *people* have assessed as an external destabilizing threat. The primer of the objective conferred upon the caliphate.

---

[25] Libya Envoy Ask UN for Weapons To Fight IS Calling for Respect of Elected Government – The Tripoli Post 3/5//15

To establish such is not to meander, grumble in the dust in anguish, confer dearth matters in and of outsiders for resolve, but assault the scourge with decisive measures and ignore heinous contradiction, hypocrisies befitting a similar response. Outcome will render a victory or balance through negotiation, brokering. It cannot be ignored opposing forces throughout conflict are collaborating one of another. This is not war but sinister scheme. Such contradicting actions propose a planning of phase-stage initiative of sporadic actors to bring about a desired result.

Stage/Phase Interpretive Assessment

Sectarianism is an adopted war strategy parlaying and outline for future developments. Four years of Arab Spring initiatives finds Yemeni conflict (Syria, Egypt, Libya, Iraq destabilized) at center stage. Chronological short narrative detail a similitude of events confirming strategic action affected to bring about course directive. As there were many events leading to civil unrest none more pivotal as President Saleh, 2011 February 3, announcement he would not run for reelection nor pass power to his son during anti-government protest in Sana'a.

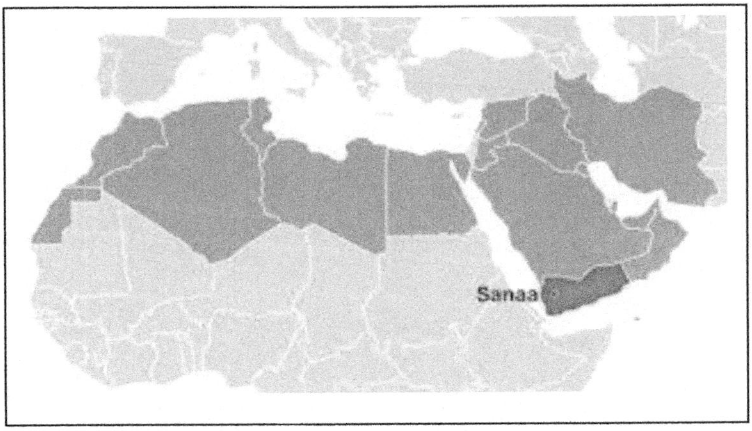

Ruling class abdicating power creates a vacuum and draws numerous opponents (both anti and pro) into the political arena of succession.

a. On February 27, Abdul Malik al-Houthi announced support of pro-democracy protest in an effort to affect regime change.

b. Precipitous action precludes the denouncement of the former regime and sets the stage for dramatic events to achieve provincial autonomy. March 19, Houthi fighters engaged pro-government force in battle seizing control of the city and forcing the local governor to flee.

c. Conflict is inevitable in matters of the power vacuum conferring a response to prior administrative authority influencing decision able to obstruct, counteract a newly established administration. Houthi formed their own administration March 26, in the Saada governorate independent of Yemeni authorities.

d. Establishment of newly formed administrative body announces the conference of a new governing body distinct, dissolve of/from the prior ruling authority. During the months of July-August 2011, several sectarian clashes occur between Islah-Houthi, Sunni-Houthi affecting sever casualties.

e. Conflict is inevitable in the struggle for dominance and rule; it can also demonstrate one's force capability. Intense fight November 9, Houthi overrun pro-government forces Kashir and Aahm tribes and seize Kuhlan Ash Sharaf district advancing toward the port of

Midi, gaining access to the sea and strategic advantage to launch assault on the capital Sana'a.

f. Geographical terrain must be seized, contained, for the security, protection of the cultural host body. November 23, former President Saleh signs power transfer agreement in Riyadh in exchange for immunity from prosecution which is rejected by the protester and Houthis.

g. Matters of détente are remedial in securing some course of peaceful transition. Presidential election held 21 February 2012, with Abd Rabbuh Mansur al-Hadi conferred 99.8% of the vote.

h. Electoral public demonstration provides the impetus for cultural participation, approval. 26 February, intense sectarian conflict in Hajjah governorate continues between Houthi and Sunni tribesman bringing the total killed between November 2011 – April 2012 to 600.

i. Historical sanctity, significance of turbulent events are conferred upon the casualties, fatalities. On March 8, military commander and bodyguards killed by Houthi gunmen.

j. Purging of previous ranking authority is a prime initiative. Houthi sponsor Anti-US protest in September and October expanding their influence, control in Sana'a governorate other areas around the capital.

k. Reinforcement of one's political position affirmation of influence. On 18 August 2014, Houthis protest in Sana'a against fuel price increase.

l. Rebuking economic turmoil assigns legitimacy to political stance. On 21 September Houthi take control of Sana'a; prime minister resigns, Houthi sign deal with other political parties to establish unified government.

m. A political attempt to reunify the fractured state. 9 November, new government sworn in. Houthi announce they would not take part in procession; accuse Yemen's President Hadi of arming Al-Qaeda to create security crisis.

n. Political party rejects current state of affairs due to apparent compromise. January 2015, Houthi attack presidential palace, private residence seize control of both.

o. Assault is political coup to ouster current governing body. On 22 January, President Hadi and his minister resign.

p. Coup is successful. On 6 February, Houthi declare themselves in total control of the government, dissolving parliament.

q. Political party has established itself as ruling class of state.

Intervention: Sectarian Proliferation

Dramatization of Yemini conflict is post-script of calculated assessment. Coup was inevitable, orchestrated by the Western-NATO bloc international security initiative 'Responsibility to Protect' (R2P) humanitarian intervention. Saudi coalition (Operation Decisive Storm) of neighboring forces intensify the conflict whereby the coalition (Egypt, Jordan, Morocco, Sudan, Kuwait, Qatar, Bahrain, UAE) are entrenched in various stages of sectarian conflict to the point, exacerbation i.e. will lead to insurgency. Arab Spring events have provided reoccurring evidence.

Sectarian war strategy has once again provided psycho-political ploy to incite hostility, resentment throughout the aligning region. What is evident in and of the various factions is the requirement of unified Islamic state. The objective however is improbable of cow-towing to various outside interest set on perpetuating unrest. Coalition actions will at best be scrutinized of the various factions submerged in the coalition states. War strategist objective is to promote perpetual pandemic profit; geo-politically, conventionally, socio-economically conferring a foothold on matters of interest. Saudi Arabia interest in Yemen is vast. Geo-political assessment confers a possible partitioning or annexation and control of regional waterways. Unrest and approximation levies a bold argument for intervention.

Conventional assessment necessitates the need to affect security measures deem prudent to deter hostile actions that might flow to, over its borders. Socio-economically, conflict has rendered the state and humanitarian crisis that intervention has assisted in creating. This is where dire matter creates the flow of refugee and insurgency. These are calculated acts demonstrative of historical event. A divisive cultural element has destabilized the whole of the Arab-Muslim region.

Strategic Objective: Realignment

Much can be opined of the events spearheading orchestrated regime change by Western-NATO powers in concert with Arab nations. Beholden of the surge is the fact many of the neighboring states are experiencing human rights violations (hypocritically) governed by similar states of despotic rule. The plan initiated calls for a broad cultivation of unified governance across a geopolitical 'expanse' allowing for external directive. "The term 'expanse' is treated as a "politically arranged soil"...German political economist Friedrich List formulated an economic autarchy conception of *large spaces*...in the form of a suggested customs union to compete with the world."[26] Observation holds Libya, Syria, and Egypt hold ripe

[26] The Post-Soviet Expanse: Ideological Aspects of Geopolitical Expansion Central Asia and the Caucasus Vol. 11 Issue-2 2010

strategic importance in the 'annexation' by Western-NATO powers as their geographical location provide exclusive one-way access via Mediterranean and they are well resourced.

Economics

Deprive, affect the sanction, embargo, blockade

Deprivation

Former President Bill Clinton at a world conference once opined, 'poverty is man-made,' as in many conditions of war. Assault and aggression can take the form of economy. Deprivation, various elements thereof (discrimination, ostracism, apartheid, preference, concession) to name a few, affected broadly produce consternation pertaining to matters of necessity. It is a weapon utilized in every aspect of the psycho-socio economic dynamic of exchange.

Prosperity speaks of progress, and there without speaks of the latter. Of ancient, conquers lay siege to fortified city in a manner to 'deprive' the inhabitants of necessities, to in effect 'starve the inhabitants out,' increasing the chance of surrender. This type of aggression is imposing of deprivation. Although barbaric war strategy prove quite effective and has been reworked of modern day civilization in the form of 'sanction' and 'embargo.' These two types of war strategy subscribe to those directed a culmination of the list of disparities wrought of deprivation.

Sanction

War strategy is intuitive of alternative method espousing armed conflict. During the Woodrow Wilson administration the former president assessed, "*sanctions* could be utilized as a deadly force, very effective diplomatic tool."[27] "Article 41 of the United Nations Charter authorizes the Security Council to decide on [measure not involving the use of armed force]...these may include complete or partial interruption of economic relations...sanctions can be ordered by a regional organization...imposed unilaterally by one country against another...may be imposed on moral, political term (as demonstrated against the South Africa apartheid regime)...supporters may even justify them as alternative to military action."[28]

Sanctions parlay a devastating effect per the equation $\sqrt{d + d^2} = s$. Destabilization is a primal element of deprivation disrupting the routine performance of daily operations in such ways as to cause broad socio-economic collapse. Prime example is the economic effects (cost) of the Arab Spring

---

[27] Sanctions – Elliott, Hufbauer, Oegg: Library of Economic Liberty www.econlib.org/Enc/Sanctions

[28] Crimes of War: Sanctions T. Gjelten www.crimesofwar.org

uprising devastating regional North Africa, Middle East productive output. By the fourth quarter of 2013 the estimated lost output of production was 800 billion. HSBC bank forecast assessed, "the combination of severe fiscal deterioration, and a decline in government effectiveness, security and the rule of law will weigh heavily on policymakers' efforts to bring employment back to pre-revolution levels."[29]

The array of events (protest, political, civil unrest) has brought about cultural economic effect conducive of levying internal *sanctions* initiated of and by the cultural populace, self-inflicting instability on the cultural mainframe. It can be said, 'change comes with sacrifice;' evident of divisive posturing. Although much of what has occurred in the Arab Spring initiative can be assessed as a matter of self-infliction of which external forces have orchestrated, exacerbated perniciously irresolvable. The imposed sanctions (by ad-hoc coalition) throughout the North African, Middle East region have rendered economies affected to the point of collapse.

In the case of Libya '*economic*' sanctions were affected against specific individuals (namely Col. Qaddaffi, cabinet ministers, military personnel,

---

[29] Arab Spring Cost Middle East 800 billion, HSBC Estimates 10/9/13 – www.reuters.com

family members) institutions, organizations, financial, resource transactions under government control to block access thereto disabling the regime ability to operate. Executive Order 13566, issued "Blocking Property and Prohibiting Certain Transactions Related to Libya" was directed of enforcers to identify, persons, entities named of the sanctions blocking access to all [their] property interest pertaining to the Qaddaffi regime.

*The U.S. Treasury Department's Office of Foreign Asset Control (OFAC) primary enforcer, responsible for implementing comprehensive sanctions program capable of "lifting sanctions to provide meaningful support to 'rebel' groups...*This allowed the administration (in their view) to implement, effectively maintain sanction prohibition making for allowances of stage, phase ground events during conflict. The "Libya Order," as the executive order was called, froze 30 billion in Libyan assets held by the United States. General licenses were require to conduct various business transactions (on an international scale) which in themselves led to inconsistencies between countries imposing sanctions. This curtailed Libya's ability to conduct business (preventing timely compensation of military, security forces, forestalling acquisition of food and medicine) provide services, creating hardship within the civilian population to the point of consternation. Blocking the personal assets of Col. Qaddaffi and the government via sanctions resulted in the isolation and containment of the regime allowing for the leader himself to become vulnerable to compromised logistics which led ultimately to assassination.

Demonstratively, the Libya exhibition is expressive of the ultimatum directed Syria. United States issued three (3) executive orders pertain to sanctioning the government of Syria...*Executive Order 13572 prohibited transaction with Syrian officials, two entities (namely Syrian General Directorate and Islamic Revolutionary Guard. Executive Order 13573 levied sanctions against President Bashar al-Assad, his Vice President, Prime Minister, Ministers of Interior and Defense, Chief of Military Intelligence, and the Director of Political Security, unnamed individuals and entities. Executive Order 13582 imposed the effect of an 'embargo' on Syrian nation. The order prohibited U.S. persons:*

- Any transaction involving the Government of Syria, its agencies, instrumentalities, or controlled entities;
- New investments in Syria;
- Export direct, indirect supply of services to Syria;

- Import of Syrian-origin petroleum and or products;
- Any transaction dealing related to petroleum and or products of Syrian origin; and
- Facilitation of otherwise prohibited transactions by foreign persons."[30]

Clearly to grasp the 'sanction effect' of an 'embargo' an historical analysis.

Embargo

"The embargo is the perfect example used by anti-Americans everywhere to expose the hypocrisy of a superpower that punishes a small island while cozying to dictators elsewhere."

-Moises Naim–

'Cuba,' if referred to by any other name would be called, 'embargo.' "First imposed by the United States on Cuba 19 October 1960 (almost two years after the Batista regime was deposed by the Cuban revolution)... embargo was placed on exports...after Cuba nationalized American-owned Cuban oil refineries...on 7 February 1962, embargo

[30] The Arab Spring and OFAC's Targeted Sanctions – Lord, Georgi, Burton of Arent Fox LLP; Munro – Societe Generale Americas Pg. 46-47

was extended to almost all imports."[31] Various legislation, Helms Burton Act 1996 [which penalized foreign companies that do business in Cuba by preventing them from doing business in the U.S.], Cuban democracy Act of 1992) has been enacted to increase pressure (with very little result) on the Castro administration.

Strategic Effect-Outcome Similarities

Although sanctions are considered alternative to military action, from the observation, alone they create pandemonium, dissension and unrest without the assistant of the initiator of the sanction. The most devastating effects produce a discontinuity within the governing body as to bring about internal conflicts ongoing as such within Libya, Egypt, Syria and Yemen. Geopolitical diplomacy of deprivation action provides caustic effect as directed at the host populace; the civilian.

It is not to say the suffering of the people not horrific of itself, but the repercussions of such atrocity find in future parlance to past events (always it seem) the birth of a prosperous new beginning.

---

[31] United States Embargo Against Cuba (Excerpts) – wikipedia.org

Navigating Export Sanctions

Historically, Cuba provides the litmus for bold assessment of an embargo allowing a contemporaneous evaluation of similar sanctioned action imposed (a culmination of deleterious effects) on the Syrian government. Sugar played an endearing role in the Cuban embargo saga of the 60's being a chief export. In the new millennium the primary resource (oil) gains speaks to the attention of the global market.

In June of 2009, Reuter report published oil was the second leading Cuban export (an economic boost via petrodollar-USD. transaction) conferring the matter to conducting international trade with third parties (export deflection) eased extraneous circumstances which might cost, hamper, increase productive output throughout duration of the continuing embargo. Demonstratively, the measure taking under enduring sanction (for the full duration) enhanced the Castro government ability to navigate the international arena of trade and politics posthumously stalemating prohibitions to the point of no effect, confirming the importance of competent export economic strategy in the geopolitical arena.

Of a similarity is the recent revelation of Syrian businessman targeted for EU sanctions described as acting "middleman" for the Assad government under embargo. The scenario placates contentious scheme of state function rather than international, although set on achieving the economic (as well political) objective. Divestment of insurgent adversary in such manner as to affect business partnership.

"ISIL fighters captured the oilfield of eastern Syria in 2013. Since then, the regime is believed to have funded the jihadist by purchasing oil from ISIL...Instead of merely being a customer for ISIL oil, the regime is understood to be running some oil and gas installations jointly with the terrorist movement. (The middleman's company), operates a gas plant in Tabqa, a town in central Syria which was captured by ISIL August (2012). Officials believe this installation I being run jointly by ISIL and personnel from the regime. The gas facility continues to supply areas of Syria controlled by Mr. Assad. Other oil and gas fields in ISIL's hands are thought to be operated by personnel who remain on the payroll of the regime's oil ministry. The oil is then sold to Mr. Assad, who distributes it in areas he controls at relatively low price, helping him to win the loyalty of the people."[32]

This is socio-political-economic war strategy unhindered of imposed sanction rendering diplomatic assault ineffective and conferring (by internal observation) an administrative competence

---

[32] REVEALED: The Oil Middleman Between The Syria Regime and ISIS – www.businessinsider.com – 5/7/15

on those sanctioned there directed. Whether state or international, island or land-locked the implication of embargo confers a determination on those there victimized thereby. The imposition of such is an act of war. Other effect-outcome similarities make for disconcerting argument.

Property Claims and Seizures

Sticking point of the Cuban embargo contributory of Castro seizure of American property through the process of nationalization. These American owned assets, *at the implementation of sanctions* valuated at 1.8 billion have accrued interest which translates in value (today's market assessment) to 7 billion in claims. The Helm-Burton Act expresses the requirement of a possible 'reconciliation' with Cuba whereby settling the outstanding claim issue must be addressed and resolved. Long-standing embargo has in of itself manifested improbable resolve pertain to property claims. Cuba is unlikely to have cash on-hand reserves for such litigious action. Time has claim many of the claimants, some companies no longer exist. Time has afforded Cuba opportunity for the irresolute nature of trade negotiation that will forego the way of diplomacy.

"There's a scenario that I see, which is bit by bit the fundamentals of the embargo chiseled away by executive order, by the economic and family ties linking Cuba and the United States, and by non-enforcement...In that scenario, the claims might someday be resolved, but wouldn't hold the process of reintegrating the United States and Cuba hostage...

–Julia Sweig– Cuba Expert: Council of Foreign Relations

The issue of property claims will go the way of brokering and negotiation as a matter of 'reinvigorating economic policy' whereby those holding claim [by the process of 50 years of mergers and acquisitions] are claimants of corporate entities who held title to the original assets...One of the largest claimants today, for example is Starwood Resorts, a company that didn't exist in 1959, but received a claim on the ITT Telegraph Tower when it acquired another company."[33] Starwood Resorts (by some assessment) may be willing to negotiate acquisition for ownership, development of beachfront property. It could be the litmus of a bold economic trade stimulating policy resolving the

[33] Cuba, You Owe Us 7 Billion – April 18, 2014 by Leon Neyfakh www.bostonglobe.com

existing claims providing Cuba bold access to the international trade arena.

In comparison, time will impact the outcome of the Syrian embargo where the dimension (insurgency, insurrection, civil war) as well matters concerning resources and assets are under constant siege, were not imposing elements in the matter of Cuba. The Assad regime would have to navigate and survive the onslaught to achieve a national reckoning.

Exiles, Refugees - Cuba

Paradise is a pronouncement, there born suggest all outside boundaries assail a moot pointlessness. Tragically enough, governance and the politics thereof occasion to disturb the balance. The American embargo imposed on Cuba exacted a price on nearly three generations of the indigenous population. Embargo traumatic effect is thoroughly examined in the historic 'excerpts' of PBS narrative, "Cuba Exiles in America."

"The Cubans who came to Miami in 1959 were supporters of the ousted Batista government...joined by increasing numbers of wealthy Cubans whose property had been confiscated [by] the [Castro] government. Many of these pioneers left Cuba with nothing and had to begin anew. Sugar mill owners became gas station attendants;

professional women took jobs as maids...Cuban exiles fought Castro repressive regime from abroad, many committed acts of terrorism...some involving the U.S. government such as 'Operation Mongoose.'

By the mid to late 60's a swell of discontent rose in Cuba, fed by economic hardship...when Castro closed down 55,000 small businesses ...more Cubans turned against the revolution. As pressure mounted, Castro opened the port of Camarioca. Relatives from Miami came to collect those left behind in Cuba. By 1974, a quarter-million Cubans had been welcomed into the United States...refugees arrived indirectly through countries such as Spain and Mexico.

Between April and September 1980, 125,000 Cubans arrived in Florida from the port of El Mariel...It all began when a bus crashed through the gates of the Peruvian Embassy in Havana. Two guards were wounded...Fidel Castro, in a fit of anger, removed the security post from the embassy entrance...within hours there were 10,000 Cubans inside the embassy and thousands more on the way. Embarrassed, the Cuban government called the refugees scoria: 'trash.' Castro decided to open the port of El Mariel to anyone who wanted to leave Cuba...

Consequence of economic hardship affected of embargo achieves the objective concerning the state and its people always resulting in political socio-economic flight. It creates the animosity and divide the adversary desires to bring about destabilizing effect. Although, in matters concerning exiles and refugees it creates a 'diplomatic socio-economic political bridge.' From hence come the pronouncement of future political reformation, change. The historical excerpt narrates formidable argument.

...Of all the aspects of the Cuban revolution, none has had a greater impact on America than the immigration of over one million Cubans to the United States...they recreated and reinterpreted Cuban culture in a new homeland...[transforming] Miami into a Latin American city...the epicenter of the Cuban exile community, built on strong Cuban coffee, Cuban food, Cuban music and Cuban business sense. But mostly it was built on politics"[34]

Out of disparity, determination and drive allowed for the cultivation of a homeland new, an annexation of occurrence which will allow ambassadors from both side familiar, *in the very near future no doubt*, to cultivate out of the desolation of a strained embargo a political socio-economic reconstruction of post-embargo Caribbean Cuba.

Syria

The objectives of embargo are clearly detrimental to all residing in its boundaries. As if not enough, civil war devastates the whole of a nation's economic base as it culturally implodes, wreaks havoc internally on its primary asset; the citizenry. As of second quarter 2015, unemployment in Syria is a staggering 57% with the government directing a considerable amount of its budget on price controls.

33 PBS American Experience – Fidel Castro www.pbs.org/castro/peopleevents

Syrian conflict has created a steady flight of refugees to neighboring countries. "Lebanon, Turkey, Jordan and Iraq are cutting back sharply on the number of Syrians they allow on their soil...capacity of the host communities has been stretched to the limit.

Here lies a broader element of the diplomatic socio-economic political bridge having prolific effect on the aligning region either for a better or worse case scenario. One must consider the factional, sectarian cause of flight and the repercussion of this ever-so-often occurrence in the region.

As a matter of flight, nation states are all too familiar with predictable occurrence that result from the escalation of Arab Spring initiative (unrest, conflict, civil war) within Syria (as witnessed in other destabilized nation states) would cause the indigenous civilian population to flee the conflict and seek refuge in neighboring states. Reenactment of this continual-ever-so-often process of flight is a normal occurrence.

In tumultuous times, it has been an inherent predilection to practice the action of the traditional nomadic tribes. It is considered ritualistic and to some a sacred evolutionary process. Although, the polarity of politics has defiled the practice. Where the ancient were accustom to assimilate various cultural nuance from ethnic diversities migrating, refugees in flight are beholden of the bias, and stereotyping of the host nation, only embraced for and of the international aid, concession duly proffered in such crisis.

Caustic effects call into question the humanitarian significance of global power commitment to occurrences *objectively* brought to bear on the civilian population by the provocative nature of their diplomatic proselytizing. Flight from ensuing conflict proscriptive, predictable of the civilian caught in the crosshairs of civil war. The global war strategist assessment dealing with the refugee crisis of the Syrian conflict (as pertaining to other recent flights from conflict) have conferred on the matter indignation worthy of rebuke. The refugee is the breath of a nation torn. They are the record, result of occurrence, history; evidence of the tyranny exacted upon their innocence by duplicitous, sinister authority ever so indictable.

"Make no mistake: The current refugee crisis is 'global'...one whose scale is unmatched since World War II. There are 19 million people in the world who are currently (of various political conflict) refugees...and they all need to find somewhere they can live in safety...we're not just talking about numbers. We're really talking about the ways nations fail refugees...Often not always, they end up in

camps. Life in the camp is often difficult, cramped, and unsafe, with few prospects for work or education. And they are also national failures: they keep refugees from integrating into local communities and creating stable, productive new lives there. At their worse, camps can keep families stuck in limbo for generations...seeing that the camps offer them little hope or protection [refugees] seek out safety from persecution further afield, often in developed countries, particularly in Europe. The journey is often horrifying and dangerous: Many families drown crossing the Mediterranean in rickety boats...The trip is so perilous...because Western governments, wanting to discourage all forms of uncontrolled migration, have ***let it be that way*** *as a matter of deliberate policy...Western countries are experiencing...what happens when large numbers of refugees show up...squalid overcrowded camps...[or as]...in Greece, for example – or that are overtly hostile.*[35]

The refugee crisis is a '*humanitarian crisis*' proffered for the sake of controlling another's sovereignty where rhetoric posits unconscionable intent and contradiction. It has been expressed "the administration...admits that the best way to *solve* the refugee problem is to *solve* the political problem causing it... Rightly so! For the displacement was

---

[35] The Refugee Crisis: 9 Questions You Were Too Embarrassed To Ask – Fisher/Taub, 9-9-15, www.vox.com

cause directly by NGO provocateurs, the embargo, the support of internal opposition sects, support of rebel insurgency by none other than Western-NATO alliance and its Arab state affiliates for the sake of global hegemony.

The civilian population purports to intended sacrificial collateral damage to be exploited. Contradictory of quote expressed is the rhetorical phrase...our primary goal when speaking about the Syrian refugees is to get them home again."[36] A collaborative effort involving the United States and Turkey proposed the [creation] of a 'Syrian safe zone' for the civilians (refugees) of the conflict requiring the annexation or partitioning of a provincial area within Syria. To do so would be by international law 'occupation', hence an 'invasive maneuver/initiative.' This objective, though it be attained would classify the whole of Syria an international war zone.

Delusion assigns a pernicious cadence to the rhetoric and speaks of 'a mental state demented, conferring on the global power elite addressing the refugee issue, 'a pronounced psychosis,' where malevolence can be descriptive of arrogant political posturing. As with sanction imposed to cause

---

[36] Critics: U.S. Reaps Harvest Of Its Inertia – Charles Hoskinson, washingtonexaminer.com 9/14/15

deprivation the repercussion of such action always speak of future opportunity. Even in Syria. This is the moment subversive elements manifest into blowback.

The very nations whom created the refugee crisis are being set there upon by the hordes of displaced civilians whose homes were destroyed for the sake of an imposed democratic dictate. Regime change in the matter of Syria, as well Iraq and Libya has been globally evidence as a purposeful wave of utter destruction i.e. to cause the devastation, deprivation that results, providing the Syrian population (as well other in the region of similar circumstance) with a weapon more powerful than bombs and bullets; a dispelling argument *in evidence* of the fallacious need of democracy.

Force-feeding democratic ideal unscrupulously has rendered the region defenseless to divisive governance and rogue factions. Western-NATO power and their affiliates have deprived civilian-refugees of the basic necessities unwarranted by diplomatic, military action; violation of international humanitarian law. This contradiction of democratic intervention validates the claim of the refugee for recourse and compensation. The refugee has but to continually state cause of present conditions to the world to obtain resolve. Their plight will cause the cultural identity to be absorbed, infused as a matter of resolve within the cultural mainframe of the Western-European bloc.

Blockade

Unlike other forms of sanction, none escalate the probability of war affecting the utility of conventional weaponry to deprive a rival. Display of military armament to prevent use, access of neutral territory (land, sea, air) purports to, 'gun-in-hand diplomacy' and speaks of imminent and direct action; lethal. It is not without note deterrent provocative of those imposing such measure and should be assessed as 'declaration of war.' To assess such posturing otherwise would be absurd.

It has been discussed of recent in the halls of intelligentsia the hierarchy of power is being challenged in and of countries assailing in the arenas of finance, industry, governance. China is considered by far 'the nation' posing considerable challenge of the dominate role, preeminent superpower of its lone rival, the United States. Noted in reference of passage for example:

"China's remarkable economic momentum, its capacity for decisive political decision motivated by clearheaded and self-centered national interest its relative freedom from debilitating external commitments, and its steadily increasing military potential coupled with the worldwide expectation that soon it will challenge America's premier global status justify ranking China just below the United States in the current international hierarchy."[37]

This 'mounting challenge' to global dominance has provoked discussion among various institution of academia proffering analysis of the 'viability of blockade.' Although posited reference confers a lack of analytical in-depth-ness to scholars 'disparate beliefs,' action unanimously are considered, from an historical or analytical context as aggressive. The premise of its purpose is deprivation incurring fatalistic consequence of response, irresponsive thereto.

Strategic war tactic presents to the provocateur diplomatic challenge in of itself necessity to conduct action in a manner prudent of the objectives to avoid mishap. Current scenario assails China's military modernization (A2/AD) complex encompassing sea borders, "presents a serious and

[37] Great Power – Wikipedia – 'Hierarchy of Great Power' wikipedia.org/wiki/great power

sustained challenge to the United States [operational access] to the region...arguably attenuates the United States ability to defend its interest."[38] In this context, inference is most likely interpreted as interference or obstruction of the 'status quo' by force of progressive act; a threat.

A2/AD

Notwithstanding its regional importance, China's progression in various arenas of industry and finance has advanced its potential to produce such defensive capability countering superpower aggressions. It is not military per say but its *projection* of power to defend its territory and its neighbors, if necessary whereby superpowers realize such exhibitions can and may sway (coerce) interest from the United States in the region to embrace security measures of a close proximity. China's anti-access/area denial (A2/AD) demonstrates capabilities in design to defend and disrupt expeditionary war offenses of an aggressor imposing on its geographical periphery and territory.

"Chinese anti-access capacity includes a large ballistic missile force designed to attack key point targets,

---

[38] Stranglehold: The Context, Conduct and Consequence of an American Naval Blockade of China – Sean Mirski carnegieendowment.org/2013/02/12

such as air bases and naval facilities. Chinese area denial capabilities consist of advanced counter-maritime and counter-air systems designed to destroy critical mobile assets, such as surface ships and aircraft. A2/AD also extends into the space and cyber domains...designed to disrupt U.S. power projection."[39]

Security measures are substantial, strategically placed to address conflict disputes with U.S. regional interest. Array of weaponry (Figure 2) provides the impetus for the argument for strategic reassessment. Chinese land attack and cruise missile capability threaten U.S. air and naval bases of Okinawa and Guam. If disabled, would considerably disrupt U.S. combat force operation. Less demonstration, exhibition of such security measures has caused superpowers to reevaluate defense strategies, conventionally and psychologically. The coercive nature of military power allows for the negotiation of alliances more favorably conducive of regional cooperation. The subversive aspect allows for compromising one's extensional influence within the region. This dichotomy has broad implications necessary to achieve the unconventional dominance and is in fact prerequisite of the war objective.

---

[39] Anti-Access/Area Denial: the Evolution of Modern Warfare Major Christopher J. McCarthy, U.S. Air Force

China Anti-Access/Area Denial Weapon List*

1. SRBM/MRBM short/medium range ballistic missile

2. ASBM/ASCM anti-ship ballistic cruise missile

3. DF-21D ASBM anti-ship ballistic missile

4. S300 air defense system

5. Nuclear-powered submarine class *(operational)*

   a. Type 091 Han-class SSN 1st generation
   b. Type 093 Shang-class SSN 2nd generation
   c. Type 094 Jin-class SSBN

Figure 2 *logistical resources deployed

Blockade: Counter…measure, action, balance?

Countering another's progress with aggression portends to preventive envious digression. Action, device, force of considerable measure are determine within the strategic context to affect deterrence of objective practices universal. Absconding from the 'commercial' pronouncement, necessitates the act of war. China's anti-access/area denial capabilities prescribe regional and foreign actors address Chinese military innovation both ad hoc and collaboratively.

Substantive elements have conferred on foreign interest, namely the United States, the requirement for reassessment of regional military expeditionary forces stationed about the Asian littoral. Anti-access/area denial A2/AD provides China with broader range of expansive defense capability enabling it to defend itself of land aggression by and through protracted war at sea strategy of those expeditionary forces. Capabilities have advanced its ability to cultivate and defend broader range of disputed costal territory (perimeter outer-island chains) exacerbating tension in and of neighboring nation states.

Re-balance?

"The United States was destined to become a 'Pacific nation' [declaration due to historical occupation] *with strong commercial interest in Asia. The United States has maintained forward military presence and unimpeded access to the region in order to overcome the "tyranny of distance..."* Aggression to 'counter' aggression proposes a balancing act of sort. The coordinate response to A2/AD has been designated a 'rebalance' of the military strategy to affect a protracted diplomatic deterrent where confrontation, dispute arise. It is descriptive of Christian Le Meire's

"fleet-in-dispersal" [strategy] "...to avoid direct confrontation with China."[40] The initiative proposes:

- Deploy new (4) littoral (costal) combat ships to Singapore.
- Assign 60 percent of U.S. Navy fleet to Pacific by 2020.
- Conduct activities, exercises, and exchanges with regional navies.
- Seek logistical basing options in Australia.
- Affect strategic concept Air-Sea Battle (ASB) designed to attack-in-depth integrated operation across air, land, sea, space, and cyberspace by three basic actions: disrupt, destroy, defeat to...
- Address power projection under A2/AD threats.
- Provide wide range of striking options.
- Affect offshore control, beyond range of A2/AD capabilities.

To be consistent with such initiative requires imperative and adequate appropriation and

[40] U.S. Japan Allied Maritime Strategy: Balancing the Rise of Maritime China – Tetsuo Kotani - Pg. 7 – CSIS April 2014

provisional assurance to regional partner thus being directly affected by actions consider provocative.

Economics: Diplomacy Coercive

Advancing the argument of the subversive war dynamic requires evidence of event and occurrence proposing resolution of dispute. Socio-economic ultimatum is not exclusive of the Chinese political process and is hypocritical of foreign and neighboring states that vigorously practice such. Economy speaks of competition by and of all things tangible, intangible.

"China has pursued a strategy in Southeast Asia that relied heavily on 'economic carrots' to increase the stake of Southeast Asian countries [in] maintaining good relations with China...China has directly used economic relations to 'compel' target countries to alter their policies..."

Withstanding various interpretation of influence, examples of posited events suggest lack of formidable debate concerning territorial right could enhance regional ties. The Scarborough Shoal incident (Chinese fishing within the Philippine 200-nautical mile 'exclusive economic zone' [accosted by] armed boarding party (assuming Philippine) in attempt to arrest, two civilian Maritime Surveillance vessels intervened. Incident precipitated into sovereignty claim dispute and 'standoff' between Philippine Coast Guard and armed China Fishery Law Enforcement Command vessel).

...Incensed by Manila's unwillingness to withdraw from the Shoal [disputed territory], China resorted to 'economic measures' to punish the Philippines for 'encroaching' on Chinese sovereignty. Chinese quarantine authorities reportedly blocked hundreds of container vans of Philippine bananas from entering Chinese ports...[quarantine]...dealt a major blow to the Philippines which exports more than 30 percent of its bananas to China. China began slowing inspections of papayas, mangoes, coconuts, and pineapples from the Philippines...Chinese travel agencies stopped sending tour groups to the Philippines...Filipino business leaders pressured the government to 'abandon' its confrontational approach in the Scarborough Shoal."[41]

Economic strong-arming is projected in many forms. Comparison will illustrate. August 2015, Chinese devaluation of the Yuan confers a deprivation on global trade.

"With the U.S. average 'tariff' on industrial goods well under 2 percent, this change in China's currency value easily 'swamps' most U.S. tariffs...

---

[41] China's Coercive Economic Diplomacy: A New and Worrying Trend – Bonnie S. Glaser – CSIS Aug. 6 2012

[In comparison]

...Ultra-low interest rates [QE] are likely to have bolstered housing prices by lowering the cost of mortgage credit...little evidence that ultra-low interest rates have boosted equity markets...housing prices and bond prices are higher...[But]...it is more difficult for today's household to borrow against any increase in wealth because of tighter credit standards...Ultra-low interest rates have prompted additional capital flows to emerging markets, particularly into their bond markets. Emerging markets that have high share of foreign ownership of their bonds and large current account deficits will be most vulnerable to capital outflows if and when central banks begin tapering current policies"[42]

Debilitating effects are felt in the undermining of the ability to acquire, increase one's wealth position against invested interest irresolute of policy dictate. This is force ultimatum descriptive of no alternative. Globalized financial markets diversify risk and confer on the 'elite' (G8) mechanisms to improved their position during economic crisis, although severity of such occurrences are experienced throughout the global community of, dire consequence meted those of emerging economies.

---

[42] QE and Ultra-Low Interest Rates: Distributional Effects and Risk, McKinsey Global Institute – Nov. 2013

Foreign aid is a type of diplomacy persuasive of intended purpose as illustrated thus:

"[Aid] given as a foreign policy bribe. This is not the same thing as aid used as a tool of public diplomacy, because its target is a foreign country's ruling elite...America gives Egypt money and in return the enriched Egyptian military, with its prestigious military weaponry, promises not to attack Israel."[43]

This manner of aid extends into the corrosive as the *'congressional funding of Syrian rebels'* illuminates the arrogance of elitist global leaders plying ultimatum by and through means contradictory of portended rhetoric. Demanding transition resonate of vain egotism. Objectives initiated atrociously by and through the exploit of the indigenous civilian population proclaiming a doctrinal of humanitarian intervention; descriptive, coercive.

None of the examples are divest of the coercive nature of diplomacy due to the fundamental practicality of the geopolitical arena in matters of commerce allowing for its dialectic propositioning throughout the global scheme.

---

[43] Does U.S. Foreign Aid Really Do Any Good? Jonathan Foreman – Washington Examiner – Sep. 28, 2015

Advancing claim toward a defensive superiority requires exaction of the noted premise of capable defiance, defense, and resolve. Exhibition is merely the contrivance by which attenuation accomplished. It is the pronouncement affecting mode of communication in matters necessitating a commencement of dialogue. Here is where the subversive action of diplomacy attempts to undermine, pervert the foundation of sovereignty.

Display of weaponry is bold threat requiring similarity in responsive action. A regional defense system is nothing more than a blockade of one's surrounding borders as oppose to an 'aggressive' weaponry shield to prevent use access of international territory, provoking hostility. War is the penultimate result of the aggressive act.

It should be noted that progress of any type outside the status quo is considered 'a threat,' whether individual, cultural, national. Controlling the whole of humanity requires, cognitive directive, resource management, security assessment of global construct and host. This is the sum game of the whole; all or nothing. Final analysis posited of a fitting conclusion.

GLOSSARY

1. **Blockade** – is an effort to deprive, cut-off access to substantive resources, communications from a particular area by force.
2. **Coercion** – intimidation of individual(s) compelling such to do some act against his or her will by use of psychological pressure, physical force, or threats.
3. **Demoralization** – to undermine the confidence or morale of; dishearten.
4. **Deprivation** – lack or denial of something deemed a necessity.
5. **Destabilization** – to cause (as a government) to be incapable of functioning or surviving.
6. **Diplomacy** – the art of dialoguing (as in international relations) with people in sensitive and tactful way.
7. **Economics** – social science which conceptualizes the factors which determine mechanics of production, distribution, and consumption of goods and services.
8. **Embargo** – the partial or complete prohibition of commerce and trade with a particular country or group of countries.
9. **Instigation** – to incite, encourage action.
10. **Intervention** – take part in something so as to prevent or alter a result or course of events.
11. **Psychology** – the study of mind and behavior which seeks to understand individuals and groups by establishing general principles.
12. **Rhetoric** – the art of effective or persuasive speaking designed to have a persuasive or impressive effect.
13. **Sanction** – a threatened penalty for disobeying a law or rule.

14. **Sectarianism** – a form of bigotry, discrimination, or hatred arising from attaching importance to perceived differences between subdivisions within a group, religion, class, faction, ect.
15. **Sociology** – scientific study of social behavior which investigates to develop a body of knowledge about social order/disorder and social change.
16. **Subversion** – to overturn or overthrow from the foundation; ruin.

Bibliography

1. Can We Trust Our 'Frenemy' Pakistan? Washington Examiner – Nov. 2, 2015
2. Revisiting China's 'String of Pearls' Strategy: Places 'with Chinese Characteristic' and their Security Implications
3. 'Experiment' assigned Afro-America with the affirmed support of academia, intellectual global community.
4. Psychological Manipulation – (excerpts) –Wikipedia
5. Complexity of Identity; Who Am I (excerpts) Beverly Daniel Tatum – www.whitepriviledgeconference.com
6. Natural selection and the Race Problem – Benjamin K. Hays
   Charlotte N.C. Charlotte Medical Journal – 1905
7. Development of Conflict in Arab Spring Libya and Syria
   From Revolution to Civil War – Maya Bhardwaj pg.79
8. Impact of U.S. War on Iraq's Civilian Infrastructure worldsocialistwebsite 11/04/02
9. Iraq: The Human Cost – webmit.edu.com
10. The Hague Journal of Diplomacy 2 (2007) 229-254 – Arguing Public Diplomacy...Foreign Policy Rhetoric
11. Preliminary to Rhetorical Democracy – Chaun-wei Hu academiaedu.com
12. Development of Conflict in Arab Spring Libya and Syria
    From Revolution to Civil War – Maya Bhardwaj
13. Pew Research – Religion and Public Life Project
    The World's Muslims: Unity Diversity - Aug. 9 2012 – pg.4
14. H.R. 4818 - pg.168
15. GAO Testimony –
16. Lies Behind the 'Humanitarian War' in Libya: There is No Evidence! - Land Destroyer Report - Oct.19 2011
17. The Vicious Schism Between Sunni and Shia – www.independent.com.uk

18. Academia Education – 'Geopolitical Expansion of the U.S.
19. Financial Times – "How Qatar Seized Control of the Syrian Revolution – May 17, 2013
20. Libya's Crude Oil Production – ycharts.com Aug. 2014
21. Development of Conflict in Arab Spring Libya and Syria: From Revolution to Civil War – Maya Bhardwaj
22. Temehu.com – Interim Transitional National Council 5/3/11 – 8/8/12
23. Cheney 94: 'Invading Baghdad Would Create Quagmire' Youtube.com: Uploaded Aug. 10, 2007
24. Libyan Army Ready to Support Egyptian Military Intervention Against ISIS – The Tripoli Post Feb. 16,2015
25. Libya Envoy ask UN for Weapons to Fight IS Calling for Respect of Elected Government – The Tripoli Post Mar 5, 2015
26. The Post-Soviet Expanse; Ideological Aspects of Geopolitical Expansion – Central Asia and the Caucasus Vol. 11 Iss.2, 2010
27. Sanctions – Elliott, Hufbauer, Oegg – Library of Economic Liberty – www.econlib.org/enc/sanc.
28. Crimes of War: Sanction – T.Gjelten – www.crimesofwar.org
29. Arab Spring Cost Middle East 800 billion, HSBC Estimates Oct. 9, 2013 – www.reuter.com
30. Arab Spring OFAC's Target Sanctions - Lord, Georgi , Burton of Arent Fox LLP; Munro – Societe General Americas pg. 4
31. United States Embargo Against Cuba (Excerpts) wiwkpedia.org
32. REVEALED: The Oil Middleman Between the Syrian Regime and ISIS – www.businessinsider.com May 7, 2015
33. Cuba, You Owe Us 7 billion – 4-18-14 – Leon Neyfakh – www.bostonglobe.com
34. PBS American Experience – Fidel Castro www.pbs.org/castro/peopleevents
35. The Refugee Crisis: Nine (9) Questions you were Embarrassed to Ask – Fisher/Taub - Sep. 9, 2015
36. Critics: U.S. Reaps Harvest of its Inertia – Charles Hoskinson, - washingtonexaminer.com Sep. 14, 2015

37. Great Power – Wikipedia – 'Hierarchy of Great Power' – wikipedia.org/wiki
38. Stranglehold: The Context, Conduct, and Consequences of an American Naval Blockade of China – Sean Mirski – carnegieendowment.org 2013/02/12
39. Anti-Access/Area Denial: The Evolution of Modern Warfare Major Christopher J. McCarthy U.S. Air Force
40. U.S. Japan Allied Maritime Strategy: Balancing the Rise of Maritime China – Tetsuo Kotani pg. 7 – CSIS, April 2014
41. China's Coercive Economic Diplomacy: A New and Worrying Trend – Bonnie S. Glaser – CSIS Aug. 6, 2012
42. QE and Ultra-Low interest Rates: Distributional Effects and Risk, McKinsey Global Institute – Nov. 2013
43. Does U.S. Foreign Aid really Do Any Good? – Jonathan Foreman – Washington Examiner – Sep. 28, 2015

Acknowledgement

The author expresses appreciation to the family of NASA for its contributions scientific, educational providing such magnificent photos of the Earth.

Made in the USA
Monee, IL
04 November 2026

67381847R10075